Easy Vegetable Garden Plans

Created and designed by
the editorial staff of
ORTHO BOOKS

Editor
Sally W. Smith

Writers
Pamela K. Peirce
Michael D. Smith

Illustrators
Deborah Cowder
Cyndie C. H. Wooley

Designer
Gary Hespenheide

Ortho Books

Publisher
Robert B. Loperena

Editorial Director
Christine Jordan

Managing Editor
Sally W. Smith

Acquisitions Editors
Robert J. Beckstrom
Michael D. Smith

Prepress Supervisor
Linda M. Bouchard

Sales & Marketing Manager
David C. José

Graphics Coordinator
Sally J. French

Publisher's Assistant
Joni Christiansen

Address all inquiries to
Ortho Books
Box 5006
San Ramon, CA 94583-0906

1 2 3 4 5 6 7 8 9
97 98 99 2000 01 02

ISBN 0-89721-287-8
Library of Congress Catalog Card
Number 95-74577

THE SOLARIS GROUP
2527 Camino Ramon
San Ramon, CA 94583-0906

Acknowledgments

Consultants
Cass Peterson
Paula Ann Winchester

Editorial Coordinator
Cass Dempsey

Copyeditor
Rebecca Pepper

Proofreader
David Sweet

Indexer
Frances Bowles

Special Thanks to
August Broucaret
Deborah Cowder
Saxon Holt
David Van Ness

Color Separations by
Color Tech Corp.

Printed in the USA by
Banta Book Group

Photographers
Names of photographers are followed by the page numbers on which their work appears.
R = right, C = center, L = left, T = top, B = bottom.

John Blaustein: 85T, 85B
Patricia Bruno/Positive Images: 86
Wendy W. Cortesi: 3B, 10B, 80–81, 84T, back cover TL, back cover BR
Crandall & Crandall: 88T
David Goldberg: 82
Saxon Holt: Front cover, 7, 84B, 87, back cover BL
Dwight Kuhn: 1
Michael Landis: 10T, 13
Ivan Massar/Positive Images: 88B
Ortho Photo Library: 3T, 3C, 4–5, 8–9, 12T, back cover TR
William Reasons: 12B
Cheryl R. Richter: 89

Front Cover
Most of the gardens in this book are adapted to elegant raised beds like this, which allows the gardener to tend the bed sitting down.

Title Page
An abundant harvest is only one of the pleasures of raising your own vegetables—but it is a substantial one!

Back Cover
Top left: Planting seeds in a raised bed.
Top right: A raised bed made with railroad ties.
Bottom left: A porous hose watering system.
Bottom right: 'Golden Pearl' tomatoes.

Easy Vegetable Garden Plans

EASY GARDENING

Old-fashioned gardening was a lot of work. This chapter gives you pointers for saving time and effort in the garden by using modern techniques.

5

THE GARDENS

This chapter offers 10 garden plans plus a multitude of variations—a vegetable garden to suit every site and every gardener's taste.

9

VEGETABLE GARDENING TIPS

In addition to having an easy vegetable garden, you want to have a successful one. Follow the pointers in this chapter for a healthy, bountiful garden.

81

Easy Gardening

Old-fashioned gardening was a lot of work. This chapter gives you pointers for saving time and effort in the garden by using modern techniques.

Gardening should be fun. Even if your main reason for gardening is serious—to save money or to have better or healthier food—there's no reason why you shouldn't enjoy it, too. Millions of people garden just because they like to. They like watching the tender seedlings push up through the earth, the anticipation as their plants grow, and the fun of harvesting and cooking their own fresh vegetables. As with any complex project, however, some parts of gardening are more fun than others. It makes sense to minimize the unpleasant parts so that you can spend more time on the fun parts.

If you spend too much time scraping weeds off with a hoe in the hot sun, you aren't going to find gardening to be much fun. But if you had a garden that had almost no weeds, wouldn't that help? Well, you can. You can have a garden that you can weed in about 10 minutes a week, that you never need to cultivate, and that comes pretty close to watering itself.

One of the best ways to minimize work is to plant intensively, with little space between plants. There is less garden area to water and feed, and less bare soil to grow weeds.

Today's vegetable garden doesn't look much like vegetable gardens of earlier generations. Vegetables were once planted in rows with lots of space between. Land was cheap, and space was left between rows to allow room for a horse-drawn cultivator.

Nowadays, most of us have less land, so intensive gardening makes more sense. Intensive methods result in a higher rate of production for the area. And intensive gardening saves not only space, but work. A weed-stopping mulch on the paths will stop or greatly reduce the need to weed them. In the beds, you can use a mulch that will greatly reduce weeds. And with more crops in less space, you will be preparing less soil area.

WHAT IS EASY GARDENING?

"Easy" means several things. Easy is quick: You don't spend a lot of time taking care of the garden or doing repetitive tasks. Once the garden is put in, you spend a minimum amount of time caring for it. Today's busy lifestyles don't allow much time for daily maintenance or long hours in the garden. The gardens in this book are as free of maintenance as possible.

Easy is pleasant: Most people like planting and harvesting best. These are the tasks that involve actually handling the plants, and in which the work is varied. Weeding, watering, and spraying are liked less; they are slow, repetitive, and demanding.

Easy is simple: Following the plans in this book saves laborious research and figuring. The plans tell you which plants you need and where to purchase them. They tell you when to plant, how much to plant at one time, and where to position everything.

Easy is free of problems: The plants recommended here are among the most pest- and disease-resistant available. By using them, and by following the gardening tips in the third chapter, you will avoid many of the problems that can occur in vegetable gardens.

However, "easy" doesn't mean no work. A garden that is easy to maintain is not necessarily the easiest to put in. The secret of easy gardening is to set up the garden in such a way that it doesn't develop problems. For example, a mulch will save hours of weeding over the course of a growing season. However, you do have to go to the effort of applying the mulch. There's certainly work involved in weedproofing the paths, making raised beds, and putting in a drip irrigation system. But each job will save many hours of labor during the summer, and many of them will last for many growing seasons, saving hundreds of hours over several years.

SOME BASIC PRINCIPLES

Although the plans in this book are very different from one another, they all follow some basic principles to make gardening easier. The plans themselves make planning a garden easy; each contains a plant list and a plan for planting, including a planting calendar that you can customize for your planting climate.

But beyond that, the gardens are all based on a set of principles that make any garden easy. These principles are described in the section that follows, and are incorporated into the gardening tips that make up the last chapter.

Keeping Gardening Easy

Here are a few ground rules for easy gardening. To keep your garden as easy as possible, pay attention to these principles.

Plant only as much as you want to care for
Especially if you're just a beginner at gardening, start small. A small garden is just as much fun as a large one and is much easier to care for. After you try out a small garden, add to it to make it as large as you wish. You can reduce the size of any of the garden plans in the second chapter.

Make permanent beds and permanent paths
Plan to make the beds narrow enough that the shortest gardening member of the family can reach halfway across from a path. This is between 3 and 5 feet for most people. The beds can be any length if they are narrow enough that you can step over them. Keep them under 10 feet if you have to walk around to get to the other side. Paths should be at least 18 inches wide, or wide enough for your garden cart.

Weedproof the paths Cover the paths with old rugs or boards, pave them with brick or rock, or mulch them with gravel or sawdust.

Automate the watering Install an inexpensive battery-operated timer at the faucet to water for you automatically. This allows you to water early in the morning while you sleep in, during vacations, and even when you are too busy to think of it.

Only dig the first year Do a very thorough initial soil preparation (see page 82), then keep a soil-amending mulch on the surface at all times and never step in the beds. The soil will stay loose and friable under the mulch, so you never need to dig again.

HOW TO USE THIS BOOK

The next chapter contains the heart of this book, the plans that make it unique. Browse through it and look for the garden that most appeals to you and your tastes. Once you select a garden, consider the ways in which you may want to customize it. You will want to orient your garden to get the most sun. You can make it larger or smaller. You can modify its overall shape to fit your yard. You may wish to substitute vegetables you are fond of for those you don't like so much, or to increase the space allotted to one kind of vegetable (see page 13).

Finally, read through the third chapter for tips on vegetable gardening. If you are a beginning gardener, this information will help you to avoid common gardening problems and to make a garden that is easy to care for. Even experienced gardeners may find some interesting new ideas or viewpoints.

Getting a new garden started is the hardest part. After that, every year you garden, it gets easier and easier, and more and more satisfying.

With permanent beds and permanent paths, the gardener never steps in the beds, so tilling can be reduced or eliminated. In addition, caps on the boards of the raised beds allow sit-down gardening. You may also want to consider paving the paths, so they would never need weeding and never get muddy.

The Gardens

This chapter offers 10 garden plans plus a multitude of variations—a vegetable garden to suit every site and every gardener's taste. Each includes a plant list, a garden layout plan, and—for most—a calendar. Just follow the directions and you will have the easiest, most successful vegetable garden ever!

The plans on the following pages present many ways to enjoy the bounty of a vegetable garden, each producing a different array of delicious food. Each plan is designed for a fairly specific configuration, but can be adapted to fit your circumstances (see page 13). Some of the plans include extensions, which allow for larger harvests, either for immediate consumption, or for canning or freezing. Most include a list of alternate plants.

The first, most basic plan is intended to provide fresh-picked produce from early spring through November. If you want extra to freeze or store, plant the extension garden, which contains varieties that ripen all at once. The selections for both of these gardens are vegetables popular with most American cooks.

The children's garden will help you to share the joy of vegetable gardening with children, who are often among the most enthusiastic of gardeners. Included are many tips to help children have the successful experiences that lead to a lifelong love of gardening.

The salad garden provides fresh and unusual greens—as well as reds and yellow and blues—for your salads. It includes many salad vegetables that are not available in produce markets, as well as edible flowers and some specialty vegetables for making elegant salads.

Most of the gardens in this book are planned with rectangular beds, allowing wooden edging such as that shown here, made of used railroad ties.

The classic herb garden will decorate your yard while letting you explore fully the flavors and lore associated with many popular herbs. Based on medieval garden designs, it is meant to be ornamental as well as practical.

Three of the garden plans provide ingredients for the cuisines of other lands that have become popular in this country. Your garden can have an Italian or Mexican emphasis, or you can grow crops used in various Asian countries.

Several of the plans are for people with limited gardening space. The patio garden is actually part of a patio. For those with no land at all, there are three container gardens. Once set up, a container garden can be as easy as one in the ground, and can transform a porch or a balcony into a beautiful and useful garden.

Fans of the out-of-the-ordinary will appreciate the "gee whiz" garden. It contains easy-to-grow plants whose produce is perfectly usable but surprising in size, color, or some other way. This garden is sure to elicit comment.

THE PLANT LISTS

Each garden plan specifies particular plants, which are described in an accompanying plant list. The plant lists call for specific varieties of vegetables, chosen for their vigor, flavor, disease resistance, and, often, for their ability to produce in spite of imperfect weather. While other varieties may succeed where you garden,

Top: Some unusual vegetables that gardeners can easily grow appear in produce stands only occasionally. Clockwise from the top, these are chayote, cilantro, jicama, Jerusalem artichokes, tomatillos, peanuts, rhubarb, Belgian endive, and horseradish. Bottom: These little cherry tomatoes, called 'Golden Pearl', were introduced in 1993.

it is worth seeking out the proven winners mentioned on the lists, which give the reasons for their selection. For many of the gardens, there are alternate plant lists that give you choices within a general theme. Some of the alternates are varieties that will have a better chance of bearing a crop in a longer or shorter season, or in cooler or warmer locations. The difference may be that the alternate does not require a trellis, or it may just be of interest to anyone planning a garden on a particular theme. All of this information will be helpful if you wish to choose a substitute vegetable. Note, too, that plant breeders are constantly creating new varieties, and old varieties (heirlooms) are being rediscovered and introduced into the trade, so the offerings of seed catalogs do change from year to year.

With each vegetable description is a notation, "Ready in __ days." This is the number of days it will take to grow the crop from planting to the first harvest, a period known as the "days to maturity." This is not an exact figure, as variations in weather and local conditions may affect the time required. Also note that the days-to-maturity figure doesn't include the time to grow a transplant from seed. So, for example, a 72-day tomato can be expected to ripen its first fruit 72 days after being planted out in the garden, not after being sown indoors. In addition to tomatoes, crops customarily started indoors include peppers, eggplants, tomatillos, cabbage, cauliflower, broccoli, and Brussels sprouts. More information on starting plants indoors is given on page 85.

THE PLANTING CALENDARS

Most of the plans are accompanied by a planting calendar. The calendar is correct for the Washington, D.C., area and holds for most of the middle latitudes of the United States. The farther north or south you are of this latitude, the more you will have to vary your planting times from those given in the calendar.

The most important fixed points on the calendar are the last spring frost and the first fall frost. These dates determine the frost-free season for your area. Tender plants, such as tomatoes and corn, can be grown only during this frost-free season. Cool-season plants, such as lettuce and members of the cabbage family, can be started a few weeks before the last spring frost and harvested for some time after the first fall frost, but their growing season can still be defined by the frost-free season.

Interpret each calendar for your area by asking a nursery professional the frost-free dates for your area. Then, working in the empty bar at the top of the calendar, enter the proper months for the last spring frost and first fall frost over those lines on the calendar. Reapportion the remaining months evenly between the two dates. The calendar should then work for your region.

The dates when you actually plant will vary. Use the planting calendar as a general guide, but modify it according to the weather. If you are experiencing a particularly cold, wet spring, plant later; if your spring is warm and dry, plant sooner. Don't try to interpret the planting times on the calendar too exactly.

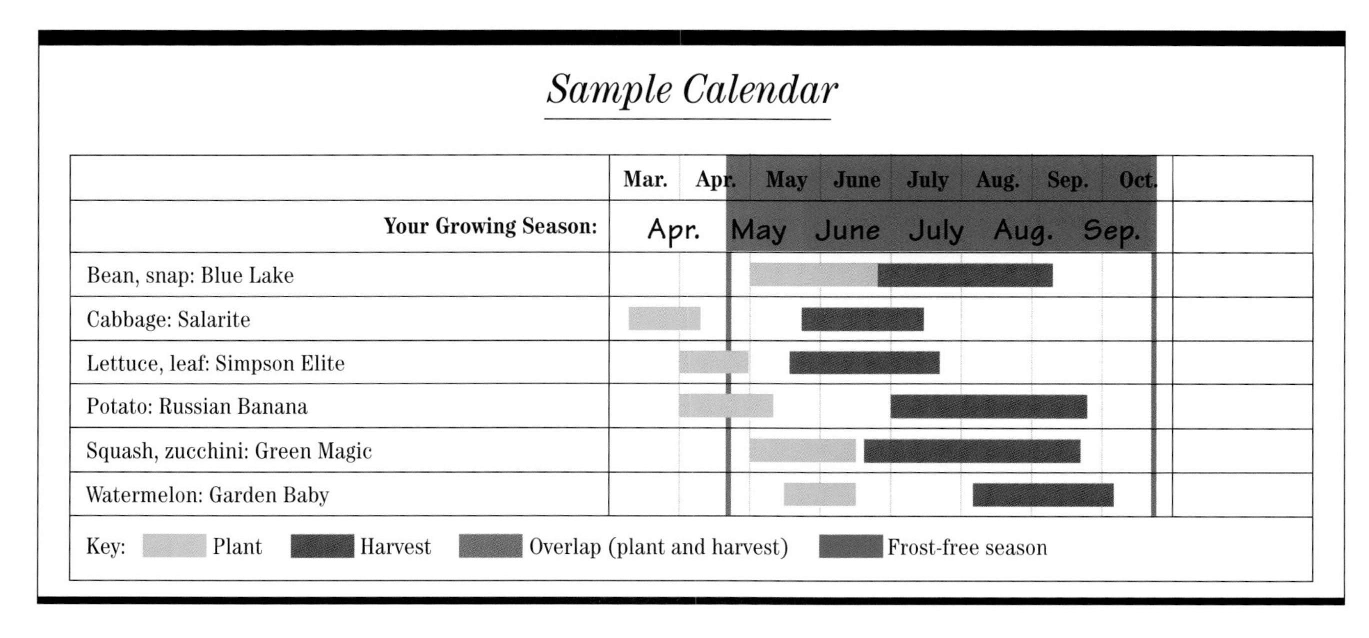

Top: The harvest is the gardener's reward. Ripe watermelon fresh from the garden is a treat enjoyed by gardeners with long, hot summers.
Bottom: Cauliflower is whiter and milder-flavored if the curd is protected from the sun. Self-blanching varieties wrap their leaves tightly around the curd to protect it—no effort is required of the gardener.

Ideal planting times last for at least a week. Read the calendar as "the first week in May" instead of "May 5."

The harvest dates—as well as the days-to-maturity information in the plant lists—also depend on many factors. These calculations represent ideal conditions; in practice, ripening often takes longer. Similarly, the calendars show the latest harvest dates for most crops as being at the end of October, but this can vary quite a bit by region. If you live in northern latitudes, once you have modified the calendar to your region, the harvest date will be accurate. In the South and on the Pacific Coast, however, the growing season extends beyond the frost-free season by weeks or even months. In USDA zones 9 to 11 (see page 91 for a map of the zones), you can garden the year around.

The calendars show an extended period for planting and harvesting some plants. For these crops, plant a small amount every two weeks during the planting period. For instance, if you plant four lettuce plants every two weeks, for most of the summer and into the fall you will be able to harvest two heads a week. This ensures a steady supply of just enough vegetables.

BUYING SEEDS AND PLANTS

Seeds are sold both from racks in garden centers and by mail order. Garden centers are more convenient and quicker, and may be less expensive. However, you may not find all the varieties you want there. To enlarge your selection of varieties, peruse the seed catalogs. Get as many as you can. Not only are they a source for many wonderful varieties, but they also contain a wealth of gardening information.

Each entry in the plant lists ends with a notation about sources for the plant. If most suppliers carry the variety, the notation is simply "widely available." You are likely to find these varieties in garden center seed racks as well as by mail order. If the variety is available from only a few suppliers, one or more of these are listed. On page 90 is a list of sources of mail-order seed, with addresses. Some mail-order seed companies have specialties, and these are noted on the list of sources.

Many vegetables can also be purchased as young plants. These are usually purchased locally, although some of the herb specialists in the list of mail-order seed sources also ship herb plants.

Vegetable seeds come in a multitude of sizes and shapes. By purchasing seeds, rather than plants, you can select from a wide variety of vegetables. If your local garden center doesn't have the varieties you are looking for, try the catalogs of mail-order seed houses (see page 90).

Certain food crops are usually or always planted from starts rather than seed. These include the perennials rhubarb, asparagus, and artichoke, which are most commonly started from root crowns.

CUSTOMIZING THE PLANS

If you wish to find a substitute for a particular plant, read the description in the plant list and look for a variety with similar characteristics, to avoid disrupting the plan. For example, a variety of summer squash might have been chosen because it is compact and takes up little room in the garden. If you substitute a large, vining variety, you will need to change the garden plan to accommodate its larger size. The easiest way to substitute is either to select a variety recommended for a different garden in this book or to choose from one of the lists of alternate plants. When you select a variety from another garden, pick up its planting calendar information at the same time.

Some of the gardens have extensions. In such cases, the first garden offered is a fairly simple one that fulfills basic needs, and the extension enlarges on that first plan.

You can modify the gardens in other ways than by substituting plants. Most of the plans are easy to modify because they are made of independent units—the permanent beds. Leave out beds or rearrange them to suit your own garden space. Remember to give each plant the same amount of space that it has in the plan, and to keep large plants to the north of shorter plants to avoid shading them.

A BASIC VEGETABLE GARDEN

This is the most common kind of vegetable garden, one designed to provide just enough fresh vegetables for your table during the growing season. The plan shown here is calculated to keep a small family supplied with all their fresh vegetables from late spring through November, with occasional surpluses for the neighbors.

Although this garden is simple, it is also the largest garden in this book. To begin with a less

Basic Garden Plan

1. Scarlet Nantes carrot
2. Jersey Giant asparagus
3. Parris Island Cos romaine lettuce
4. White Spear bunching onion
5. Red Ace beet
6. Salarite cabbage
7. Blue Lake Pole snap bean
8. Buttercup winter squash
9. Cherry Belle radish
10. King of the Garden lima bean
11. Garden Baby watermelon
12. Buttercrunch lettuce
13. Spirit pumpkin
14. Simpson Elite leaf lettuce
15. Slicemaster cucumber
16. Sweet 100 small tomato
17. Better Boy large tomato
18. Kandy Korn sweet corn
19. Little Marvel pea
20. Dusky eggplant
21. Parsley
22. Caliente hot pepper
23. Melody spinach
24. Yellow Bush Scallop squash
25. Fordhook Giant Swiss chard
26. Sugar Bon snap pea
27. North Star sweet pepper
28. Green Magic zucchini
29. Premium Crop broccoli
30. Ambrosia cantaloupe
31. Russian Banana potato

= 1 sq ft

enthusiastic garden that still provides many of the vegetables you like, eliminate beds that contain the vegetables you like least. If you remember to keep the taller plants toward the north of the garden, you can take out any beds you wish; rearrange the rest in any way you want to fit your space.

The vegetables selected for this garden plan include those that taste best when picked right from the garden and those that are expensive in stores. It includes America's favorite garden crops, the vegetables most people grow and want to put on the table, such as juicy, ripe tomatoes; fresh, sweet corn on the cob; tender early peas; and pumpkins ready for carving. For vegetable selections that appeal to gourmet or ethnic tastes, see some of the other gardens.

A few vegetables that are common in produce markets—such as celery and iceberg lettuce—are missing from this garden because, while they can be grown in home vegetable gardens, in most regions it is difficult to attain the quality you are used to.

About the Plan

The garden plan is simple, functional, and attractive. The wide central path gives room to turn a wheelbarrow or garden cart or to pile compost or mulch. Beds are 4 feet wide, making it easy to reach from a path to the center of the bed without stepping in the bed. (For a few crops, such as peas, you will need to walk in the bed to pick the far side of the row.)

The extension is based on a somewhat different premise. Most of the crops in it will need little attention while growing in the garden, but will require quite a bit of time as you harvest and process them for storage—canning, freezing, drying, or storing in a root cellar. In this large garden, the beds are longer, to save space, and are planted with large blocks of single crops. Each section will be planted only once and harvested once (or for a brief period).

Plant the asparagus first. You will do this only once; asparagus is a perennial and will bear for many years. It won't get very large the first year, so the space around it is filled with early crops that will be harvested before the asparagus needs the room.

Next, and soon, plant the peas, lettuce, onions, spinach, and other cold-tolerant crops. Plant only as much lettuce, carrots, beets, and onions as you can harvest in a two-week period. Keep planting at two-week intervals to get a continuous harvest.

A few weeks later, plant corn, and continue planting at two-week intervals. This will stretch its harvest period over several weeks. Shortly after planting the corn, you can plant the rest of the warm-weather crops.

Except for the onions and dill, the vegetables in the extension are planted later than those in the basic garden. Most ripen more or less at once, so that when a crop is ready, you can organize one large freezing project, rather than a little at a time for weeks.

The peas, however, will need to be picked every couple of days, as they become ripe. You'll have to freeze them in small batches. This snap pea variety doesn't need stringing, which saves time when freezing.

The cucumbers also need to be picked when they are ready. If the dill is ready for harvest before you are ready to pickle, you can cut dill stems and freeze them in a plastic bag.

Harvest the cabbage after the first frost, pulling up the entire plant, roots and all. Store them, roots on, in damp sand in a cellar or other cool, dark place. Under ideal conditions, this variety will keep until spring.

Plant List

Asparagus: Jersey Giant
Rust resistant, highly productive. For cold-winter areas, where the ground freezes at least a couple of inches deep. Burpee, Shepherd's.

Bean, pole, lima: King of the Garden
A very high-yielding lima bean. Ready in 88 days. Widely available.

Bean, pole, snap: Blue Lake Pole
An old favorite, still one of the best. Ready in 60 days. Pole beans bear over a longer period than bush beans. Widely available.

Beet: Red Ace
These beets develop fast and stay tender and sweet for a long time. They have good disease tolerance. Ready in 53 days. Field's, Park, Pinetree, Territorial, Vermont Bean Seed.

Broccoli: Premium Crop Hybrid
Makes good heads later into summer heat than most broccolis. Produces a large central head

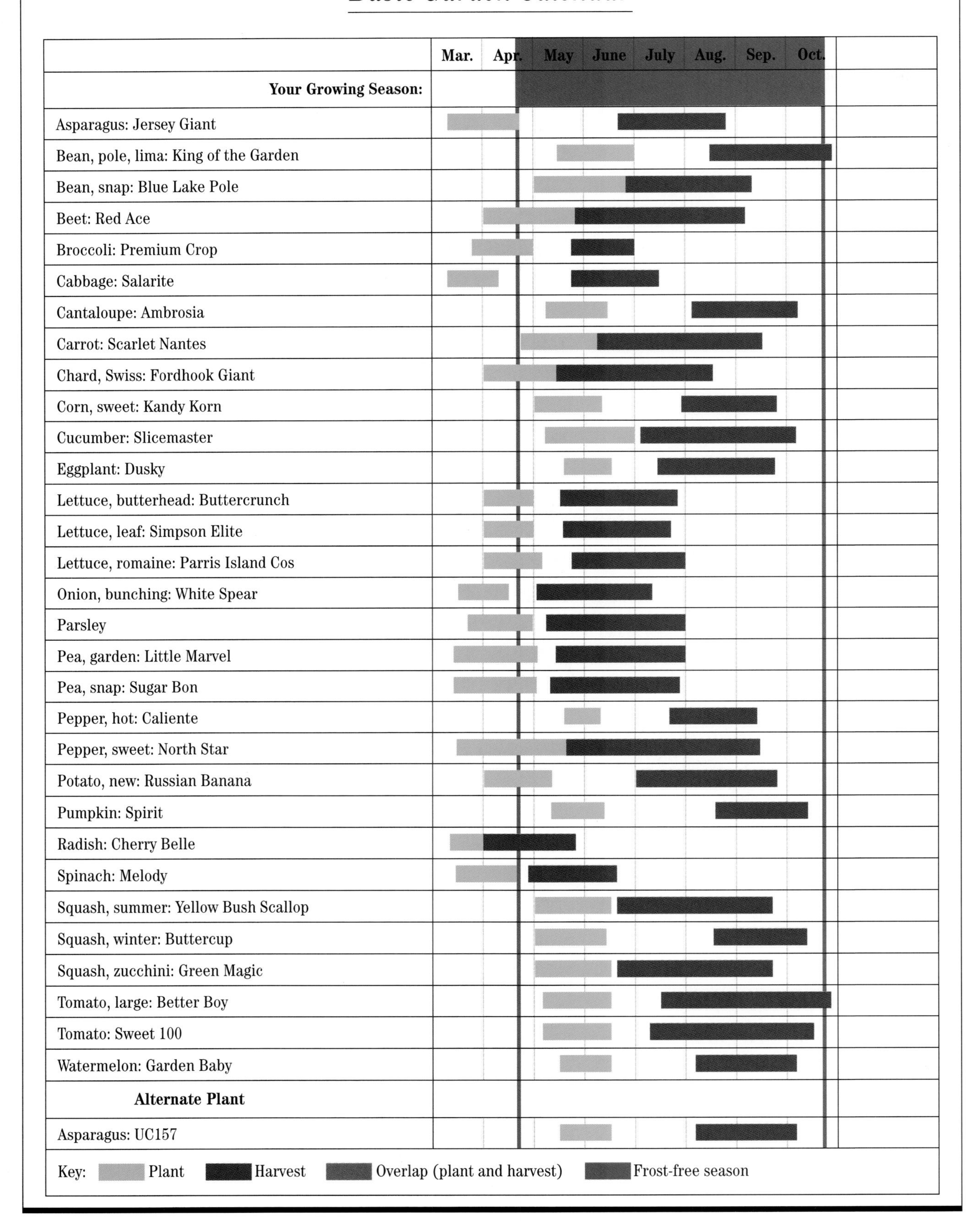
Basic Garden Calendar
Mar.
Apr.
May
June
July
Aug.
Sep.
Oct.
Your Growing Season:
Asparagus: Jersey Giant
Bean, pole, lima: King of the Garden
Bean, snap: Blue Lake Pole
Beet: Red Ace
Broccoli: Premium Crop
Cabbage: Salarite
Cantaloupe: Ambrosia
Carrot: Scarlet Nantes
Chard, Swiss: Fordhook Giant
Corn, sweet: Kandy Korn
Cucumber: Slicemaster
Eggplant: Dusky
Lettuce, butterhead: Buttercrunch
Lettuce, leaf: Simpson Elite
Lettuce, romaine: Parris Island Cos
Onion, bunching: White Spear
Parsley
Pea, garden: Little Marvel
Pea, snap: Sugar Bon
Pepper, hot: Caliente
Pepper, sweet: North Star
Potato, new: Russian Banana
Pumpkin: Spirit
Radish: Cherry Belle
Spinach: Melody
Squash, summer: Yellow Bush Scallop
Squash, winter: Buttercup
Squash, zucchini: Green Magic
Tomato, large: Better Boy
Tomato: Sweet 100
Watermelon: Garden Baby
Alternate Plant
Asparagus: UC157
Key:
Plant
Harvest
Overlap (plant and harvest)
Frost-free season

with few side sprouts, though it may make some sprouts when maturing into fall. Ready in 62 days. Widely available.

Cabbage: Salarite Hybrid
A medium-sized, disease-resistant variety. Ready in 50 to 60 days. Nichols.

Cantaloupe: Ambrosia Hybrid
Extra sweet and juicy. Disease resistant. Ready in 86 days. Widely available.

Carrot: Scarlet Nantes
Cylindrical, about 6 inches long. Ready in 70 days. Widely available.

Chard, Swiss: Fordhook Giant
Very large leaves. Ready in 60 days. Widely available.

Corn, sweet: Kandy Korn EH Hybrid
An extra-sweet early corn with tight husks that resist corn earworms. Ready in 70 days. Burpee, Gurney's, Shepherd's, Vermont Bean Seed.

Cucumber: Slicemaster Hybrid
Mild flavor, heavy yield, and multiple disease resistance make this an excellent choice. Ready in 55 days. Gurney's, Pinetree, Territorial, Vermont Bean Seed.

Eggplant: Dusky Hybrid
A black, oval variety that is early and disease resistant. Ready in 63 days. Field's, Nichols, Pinetree, Territorial, Vermont Bean Seed.

Lettuce, butterhead: Buttercrunch
A loose head that grows well in spring and summer. Ready in 64 days. Widely available.

Lettuce, leaf: Simpson Elite
Light green; tastes better in summer heat than most lettuce. Ready in 48 days. Burpee, Johnny's, Park, Pinetree, Southern Exposure.

Lettuce, romaine: Parris Island Cos
A large, green, upright romaine with white center. Ready in 57 days. Widely available.

Onion, bunching: White Spear
This variety does not make a bulb, but divides into two onions when big enough. Pick them as you need them. Ready in 65 days. Johnny's.

Parsley
The flatleaf or plainleaf types have a richer parsley flavor and are easier to chop. The curled leaf varieties are more ornamental as garnishes. It is easiest to buy plants at a local nursery.

Pea, garden: Little Marvel
An old favorite with 18-inch vines. Ready in 63 days. Widely available.

Pea, snap: Sugar Bon
Early, sweet, and disease resistant. Ready in 58 days. Widely available.

Pepper, hot: Caliente Hybrid
Medium hot, good for drying or fresh use. Ready in 65 days. Johnny's.

Pepper, sweet: North Star Hybrid
Early, for use either green or red. Ready in 66 days. Widely available.

Potato, new: Russian Banana
A small, yellow finger type for eating as a "new" potato. Mature in 90 days; "new" potatoes ready in 60 days. Johnny's.

Pumpkin: Spirit Hybrid
Compact plants bear 10- to 15-pound pumpkins that are fine for carving or pie and keep well in storage. Ready in 100 days. Widely available.

Radish: Cherry Belle
Round and bright red. Ready in 22 days. Widely available.

Spinach: Melody Hybrid
Disease resistant and dependable. Ready in 42 days. Widely available.

Squash, summer, scallop: Yellow Bush Scallop
Lemon-yellow fruits on a bush-type plant that resists squash bugs. Ready in 50 days. Field's, Gurney's.

Squash, summer, zucchini: Green Magic Hybrid
Reliably productive. Very early, medium green fruit on nearly spineless plants. Ready in 48 days. Park, Vermont Bean Seed.

Squash, winter: Buttercup
A dark green turban squash with orange flesh. Ready in 105 days. Abundant Life, Burpee, Field's, Gurney's, Southern Exposure.

Tomato, large: Better Boy Hybrid
Bears large red fruit all season on tough, disease-resistant plants. Ready in 75 days. Field's, Gurney's, Nichols, Tomato Growers Supply, Vermont Bean Seed.

Tomato, small: Sweet 100
Clusters of extra-sweet cherry tomatoes on a very large plant. Ready in 60 days. Widely available.

Watermelon: Garden Baby Hybrid
An early variety that bears high-quality round fruit on compact vines. Ready in 70 days. Johnny's, Pinetree, Territorial.

Alternate Plant

Asparagus: UC 157
An alternate to Jersey Giant for regions with mild winters. Burpee, Shepherd's, Territorial.

Plant List for Extension

Bean, bush, snap: Blue Lake 274
A bush bean that matures the whole crop at the same time, so you can freeze it all at once. Ready in 55 days. Widely available.

Beet: Lutz Winter Keeper
An old favorite that keeps well into the winter. Ready in 70 days. Widely available.

Cabbage: Storage No. 4 Hybrid
Medium sized and very dense; keeps in cold storage until spring. Ready in 75 days. Johnny's.

Carrot: Rumba
About 6 inches long. Holds well for fall and winter harvest. Ready in 72 days. Johnny's.

Cucumber: Pickalot Hybrid
Medium-sized pickling cucumbers on a semi-bush plant. Ready in 54 days. Burpee.

Dill: Dukat
Grow this type for use of leaf and seeds or for use in pickling. Widely available.

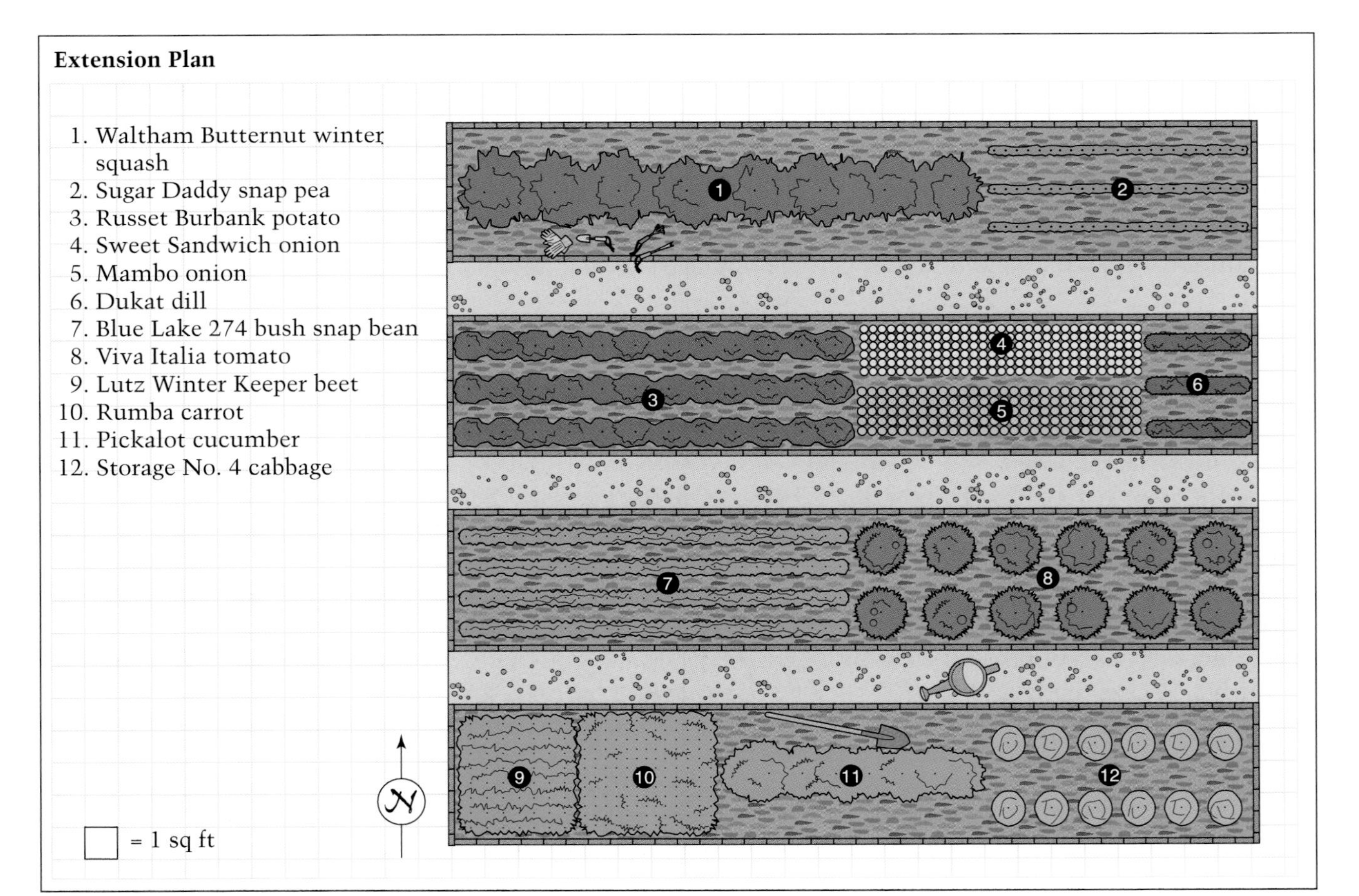

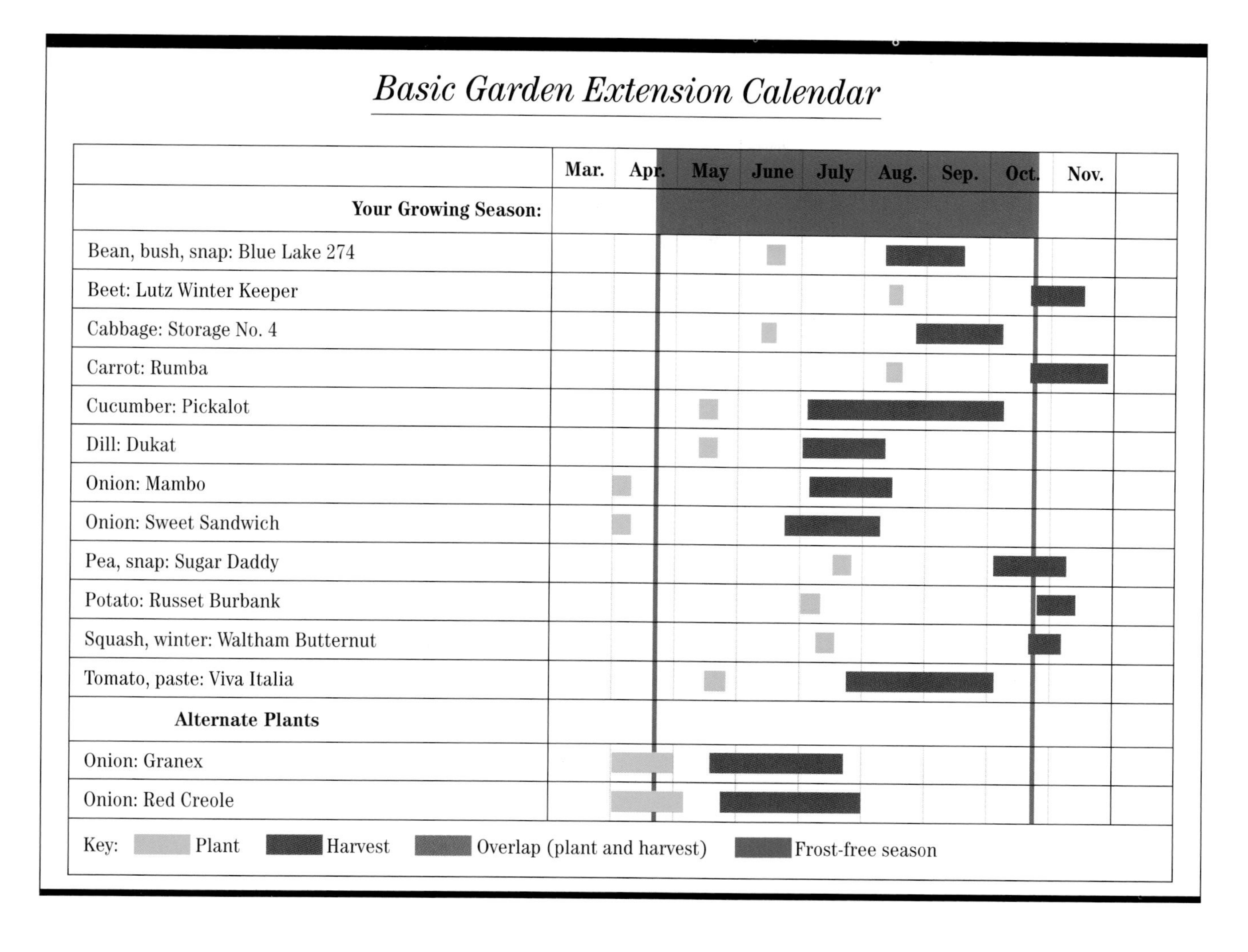

Onion: Mambo Hybrid
Large red storage onions with a medium-pungent sweet flavor. For planting in Washington, D.C., and northward. Ready in 112 days. Johnny's.

Onion: Sweet Sandwich Hybrid
Large, yellow, and keeps well. Grows sweeter in storage. Select this onion for planting in Washington, D.C., and northward. Ready in 100 to 110 days. Burpee, Pinetree.

Pea, snap: Sugar Daddy
Stringless, so it saves time when you're processing large batches for the freezer. Plants grow low enough that they need little support. Ready in 74 days. Widely available.

Potato: Russet Burbank
A disease-resistant baking potato that stores well. Ready in 120 days. Widely available.

Squash, winter: Waltham Butternut
A moderate-sized smooth yellow squash with a sweet flavor and excellent storage capability. Ready in 85 days. Widely available.

Tomato, paste: Viva Italia VFFNA Hybrid
Meaty plum tomato that ripens many fruits in a short time, for cooking. Resists a number of diseases, including bacterial speck. Needs little staking. Ready in 80 days. Widely available.

Alternate Plants for Extension

Onion: Granex Hybrid
Yellow onion derived from the variety that produces the Vidalia onion. Mild, sweet, and stores well with good air circulation. For regions south of Washington, D.C. Park, Southern Exposure.

Onion: Red Creole
Small, red, and a good keeper. Select this onion for regions south of Washington, D.C. Gurney's.

A GARDEN FOR CHILDREN

Children love to garden. The very youngest children enjoy being in the garden while things are happening. Make them a bean tepee to play in (see instructions on page 23), or plant some peas or strawberries they will be able to pick and eat. They will enjoy having their own small garden tools to help you dig and their own baskets to help with harvesting.

At about six years old, most children want to have their own garden, but they will need substantial adult supervision to be successful. By eight or ten years old, they are capable of caring for a small bed with only a little help. Some children start to think that gardening isn't "cool" when they are about twelve years old and lose interest in it. But those children who have learned to love gardening when young usually return to it after a few years.

Most children don't care whether the vegetables they grow are ones they enjoy eating. A child will get great satisfaction from presenting

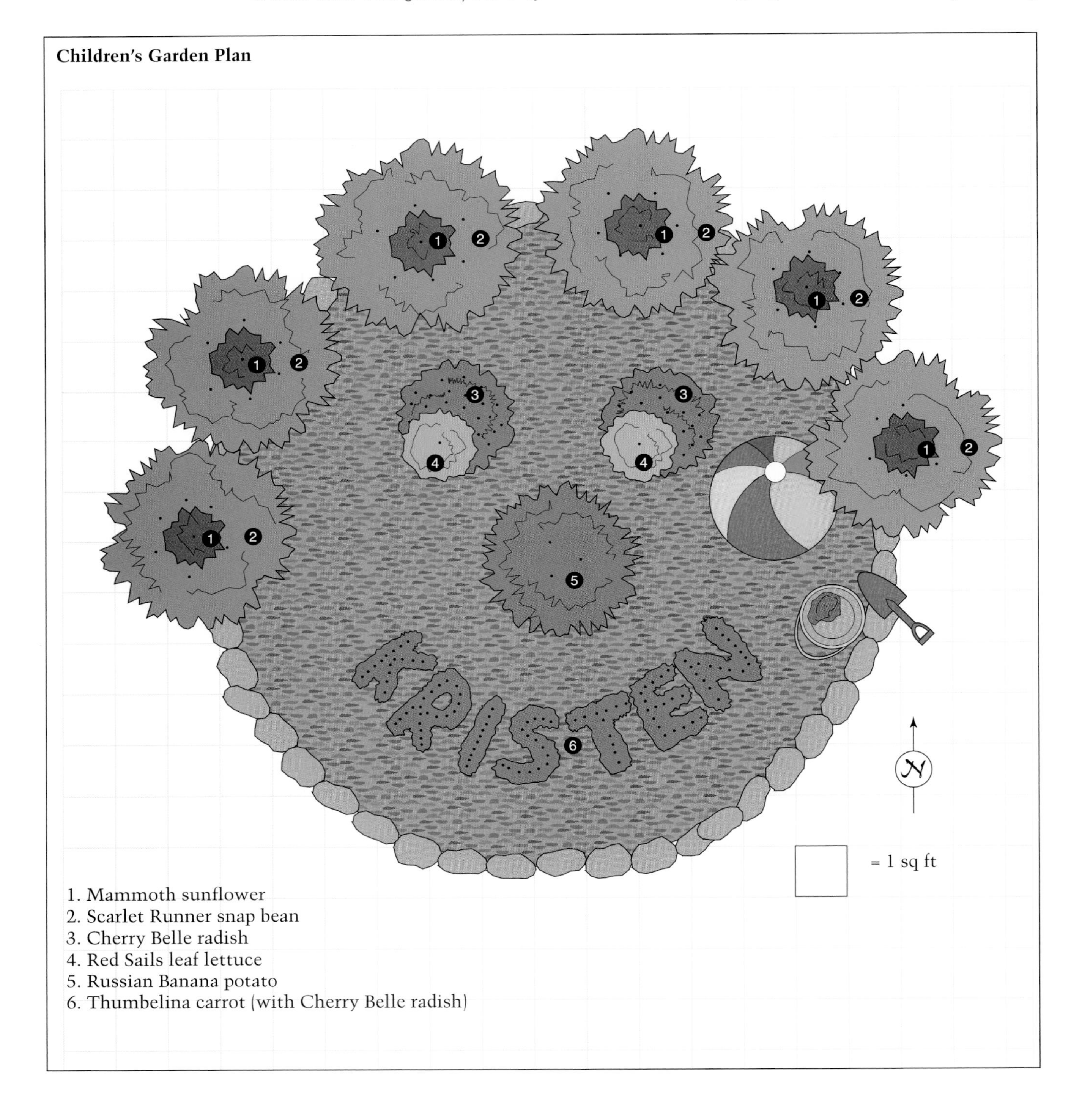

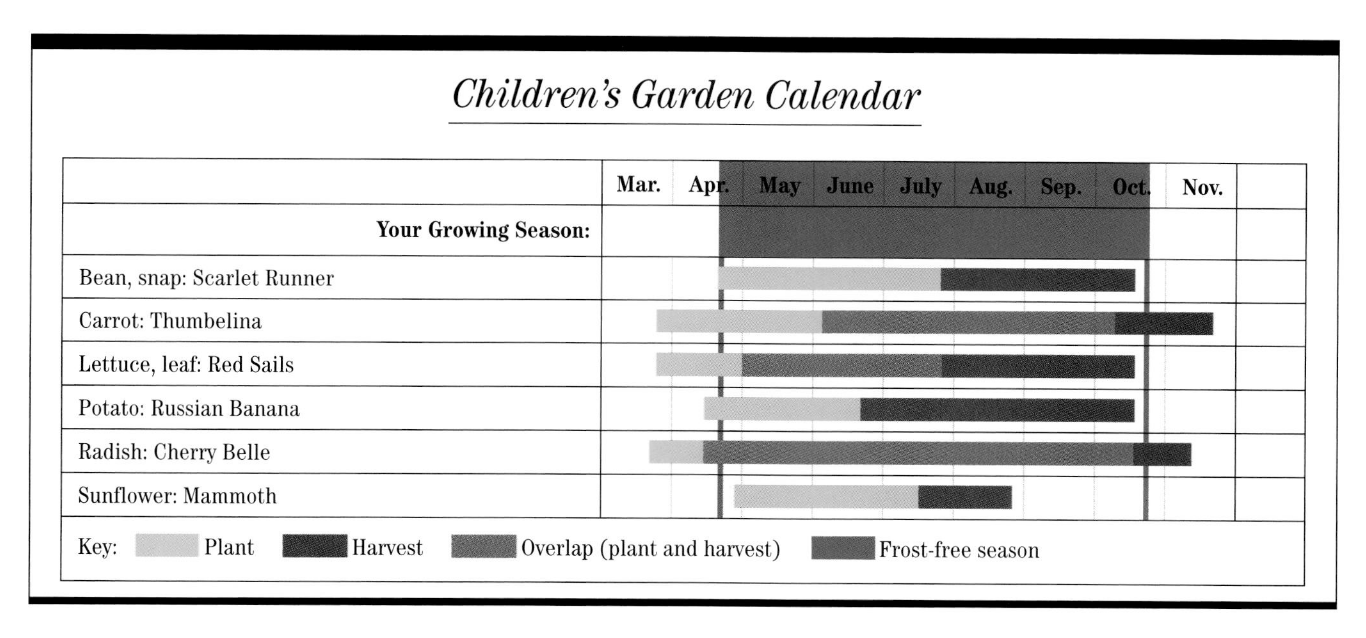

any produce to Mom or Dad, as long as parents show delight in receiving it and present it with some fanfare at the dinner table.

Make the garden to the child's scale. It should be small—2 to 4 square feet will suffice for young children. It may be part of the family garden or a separate plot just for the child.

Children like plants that appeal to their senses and spark their imaginations. They enjoy large plants like sunflowers and tiny plants like 'Thumbelina' carrots. They are pleased by unusual and funny-looking vegetables, such as kohlrabi (see page 74). They love to harvest, especially if they have to hunt for the produce. New potatoes hidden in loose soil or a mulch are a favorite. They also like making the garden pretty with flowers such as pansies or calendulas.

Children love playing with the garden; they will want to make tiny paths and edge beds with pebbles. They will want to run to the garden every morning to see if the seeds have sprouted yet. It's nice to plant radishes, which germinate in four or five days, first so they can see something happen as soon as possible.

When starting six-year-olds with their first garden, let them do as much as they can, but prepare the project in detail to ensure their success. Bring to the garden a table or sheet of plywood to work on, all the tools they will need, the seed, plant labels, and a marking pen. Have the soil turned over and loose so that they can mix in compost and fertilizer and prepare a seedbed.

About the Plan

The "smiley face" garden is designed for children just beginning to garden, between ages six and ten. For older children, extend the smiley face with some of the ideas below. Plant the carrots, radishes, and lettuce as early in the spring as you can work the soil, the potatoes a week before the last expected frost, the sunflowers just after the last expected frost, and the beans a couple of weeks after the sunflowers.

Scratch the child's name in prepared soil, about ¼ inch deep. Drop carrot seeds about an inch apart in the row, then drop radish seeds in

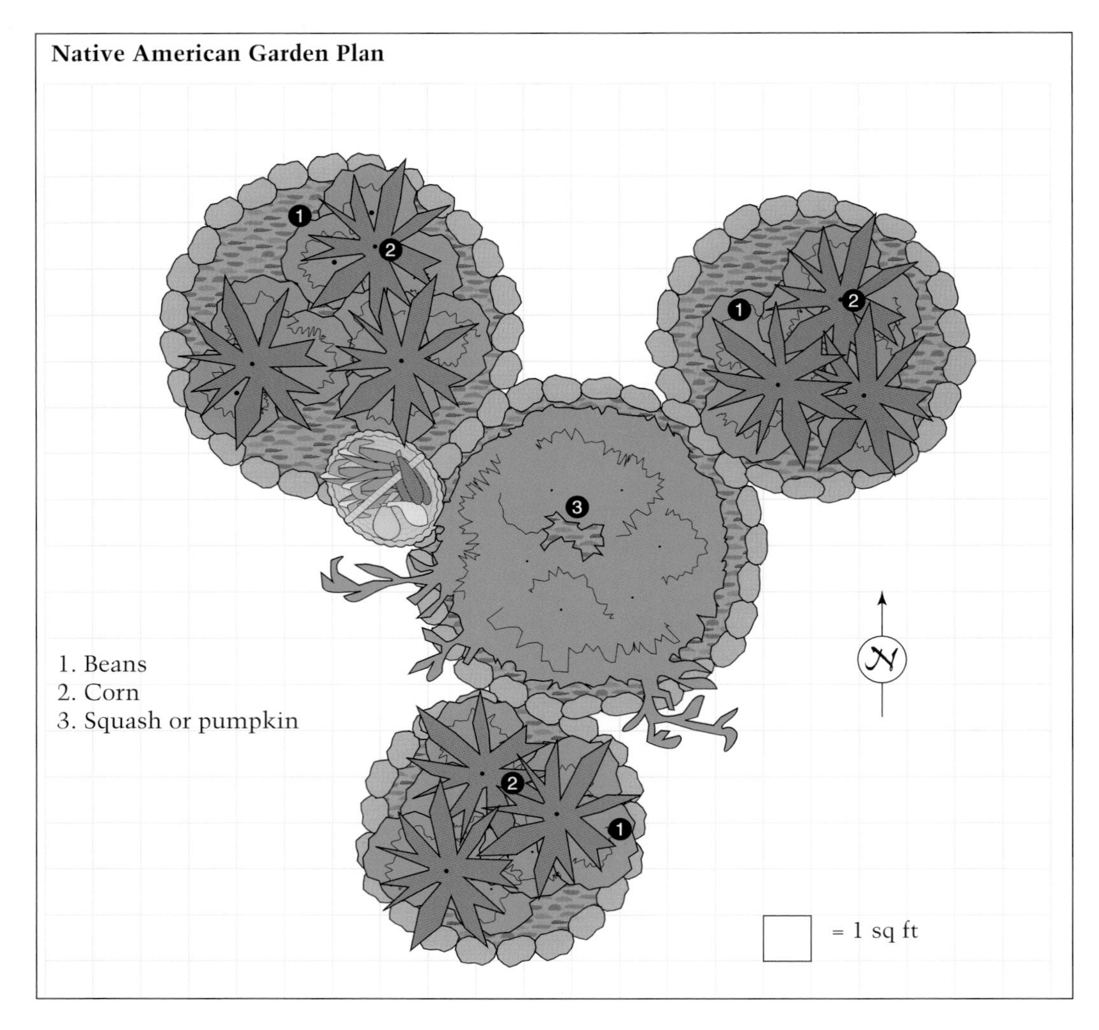

Native American Garden Plan

the same row about 6 inches apart. Cover the seeds very lightly. The radishes will come up first, marking the rows and making something to watch until the lazy carrots finally wake up (in two or three weeks). As you pull the carrots, fill each hole with compost or soil and plant another carrot in the space. This will keep the face smiling all summer.

"Plant" the three potato pieces by laying them on the surface of the prepared soil and covering them with 4 inches of sawdust or straw mulch (straw will settle to 4 inches after watering if you begin with 8 to 10 dry, loose inches). When you see the first flowers on the potato plants, begin exploring in the mulch for new potatoes, which will form between the soil and the mulch. Grope through the mulch with your fingers, being careful not to break roots. You can pick new potatoes about once a week all summer.

The sunflower "bonnet" must be to the north, or the tall sunflowers will shade the garden. The beans will climb up the sunflowers, decorating them with bright red blossoms and—later—snap beans. Keep the sunflowers well back from the smiley face; they will be *huge* by the end of summer. After harvesting the seed heads, leave the sunflowers in place as beanpoles. If you keep the beans picked, they will produce until killed by frost.

Native American Garden

For older children, plant a Native American garden. It includes three of the primary food crops of North American natives: corn, beans, and squash. Plant the corn a couple of weeks ahead of the beans and squash. This will give it a chance to gain some height before it can be overwhelmed by the vigorous growth of the beans and squash. The varieties listed first are descended from those planted by Native Americans and are available only from seed companies specializing in ethnic seeds. The alternate selections are modern varieties, more familiar and perhaps easier to grow.

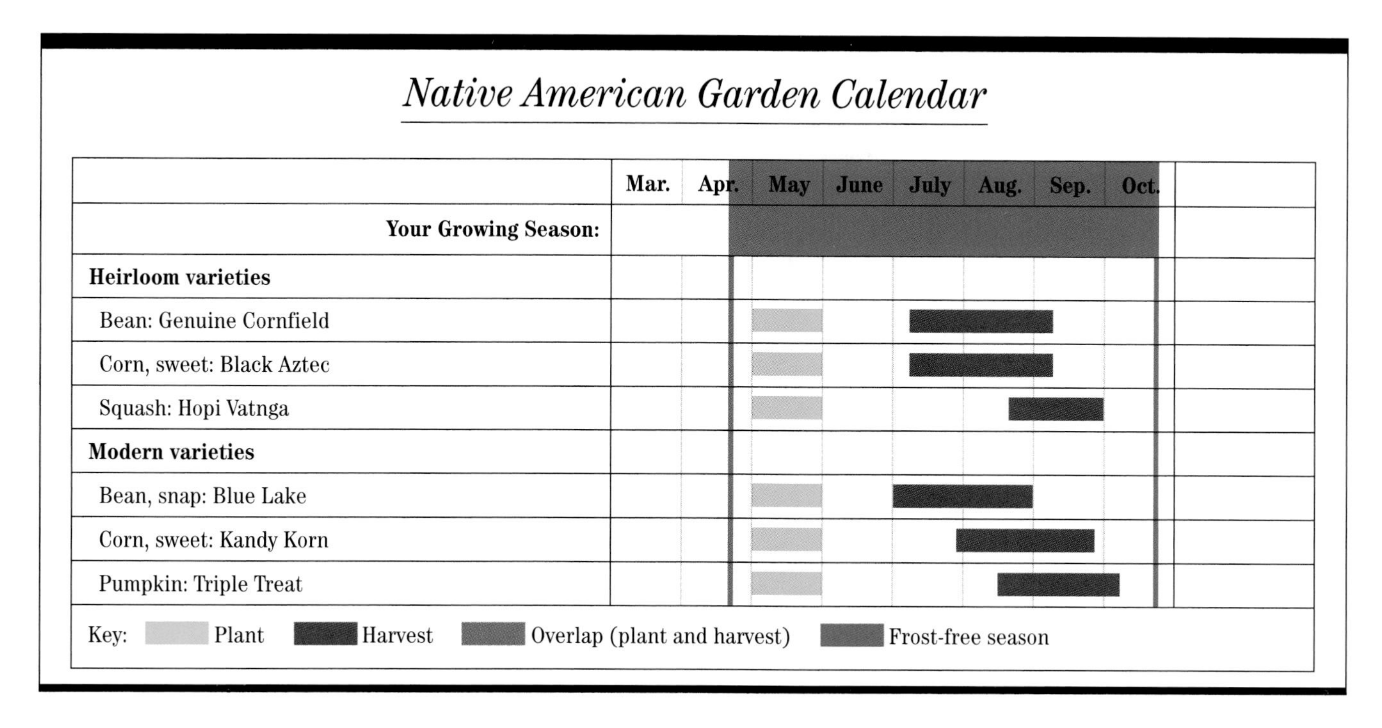

Cover the unplanted area of these gardens with a thick layer of organic mulch so that weeds won't grow in it. For extra protection, spread a thick layer of newspapers (black ink only) over the area first, water it down, and then lay the mulch on top of it, being sure to cover the edges.

Some Other Ideas for Gardening With Children

One way to extend the garden is to make a bean tepee, with 'Scarlet Runner' beans planted around the bases of poles in a tepee shape. The beans will cover the tepee in a few weeks, making a shady playhouse.

Playhouses can also be made with rings of sunflowers interplanted with 'Scarlet Runner' beans or morning glories similar to the "bonnet" for the smiley face garden. Another type of garden tent can be made of an A-frame structure of poles and string or wire on which peas or cucumbers are grown. The peas or cucumbers hang down inside, where they can be picked by children crawling into the tent.

Strange squash Any cucurbit—squash, pumpkins, melons, or cucumbers—can be grown in a bottle. Just slip the bottle over the fruit while it is still small enough to fit through the neck. Children will have a mystery to show their friends: "How did you get that in *there*?"

If the bottle is small enough, the fruit will grow to fill it, taking the shape of the bottle. The bottle can then be broken away (by an adult), leaving a bottle-shaped fruit. Using this method, children can make square watermelons, round zucchini, and cylindrical squashes. Molds for shaping squash in fanciful shapes are also commercially available. These are usually of plastic in two halves that are screwed onto the fruit, then unscrewed to release the fruit when it has filled the mold.

Signed pumpkins A child can select and personalize a Halloween jack-o'-lantern pumpkin in the summer. Scratch a child's name or initials in a pumpkin or winter squash when it is small and the skin is smooth. As the pumpkin matures, the signature will grow, making a permanent badge of identification. Faces or other pictures scratched on pumpkins will also grow with the pumpkin.

Visible germination To fill the anxious days spent waiting for seeds to germinate, children can monitor the progress of their seeds by making a germination viewer. Make a cylinder of folded paper towels inside a widemouthed glass jar. Fill the jar about halfway with moist soil, then tuck seeds between the paper and the glass at various places around the jar, an inch or two from the top. Fill the rest of the jar with soil.

Make the germination jar at the same time as you plant seeds in the garden, then set it outside so it will be subject to the same temperatures. The seeds in the jar will germinate at the same speed as those in the ground, so children can better visualize what's going on under the soil while they wait for their seedlings to surface.

Many "experiments" are possible with the germination jar. Different types of seeds can be compared, both above and below ground, during germination and early development. This is most interesting if you use very different seeds. Compare beans (a broadleaf plant) with corn (a grass), or compare lettuce (a small seed) with squash (a large seed). Or, using several jars, hold a race, watching seeds germinate at different temperatures.

Plant List

Bean, snap: Scarlet Runner
This climbing bean has red flowers and tasty green beans that are ready in 70 days. Widely available.

Carrot: Thumbelina
This small, ball-shaped early carrot is ready 60 to 70 days from planting. Widely available.

Lettuce, leaf: Red Sails
This very ornamental looseleaf has leaves splashed with dark red. Plant some every two weeks until late summer. Ready in 45 days. Widely available.

Potato: Russian Banana
A small, yellow, finger-shaped potato. Ready in 60 days. Johnny's.

Radish: Cherry Belle
A round, bright red radish. Seeds will germinate in 4 days and be ready to harvest in 22 days. You can plant more radishes in their place, or use the space for a tiny summer lettuce like 'Mignonette Green'. Widely available.

Sunflower: Mammoth
One of the biggest. If you don't like shelling sunflower seeds, put the harvested heads outside a window and watch the finches work to pull the seeds loose. Ready in 80 days. Widely available.

Plant List for Native American Garden

Heirloom varieties:

Bean, pole, snap or shelling: Genuine Cornfield
One of the oldest beans cultivated by the Iroquois. It is a shade-tolerant variety, ideal for growing with corn, and also tolerates heat well. Use for snap beans or shelling beans. Snap beans ready in 70 days. Southern Exposure.

Corn, sweet: Black Aztec
This heirloom corn has sweet white kernels at the milk stage—when sweet corn is eaten—that mature to blue-black. When dry, it can be ground into blue cornmeal. Ready in 70 to 100 days. Seeds of Change.

Squash: Hopi Vatnga
A green squash with a hard, thick shell used by the Hopi to make musical instruments. Native Seeds.

Modern varieties:

Bean, pole, snap: Blue Lake
One of the most popular home garden pole beans. Tasty and prolific. Ready in 60 days. Widely available.

Corn, sweet: Kandy Korn EH Hybrid
A large, late, extra-sweet yellow corn. Ready in 70 days. Burpee, Gurney's, Shepherd's, Vermont Bean Seed.

Pumpkin: Triple Treat
The three "treats" are jack-o'-lanterns, pies, and the hull-less seeds, which are delicious roasted. The fruit is small, about 9 inches across. Ready in 110 days. Burpee.

A SALAD GARDEN

The salad garden is primarily for those who like a distinctive tossed salad—crisp greens with a judicious mixture of other ingredients. Some of the ingredients that work well with greens in a salad also make a colorful relish plate, including radishes, scallions, snap peas, and strips of cucumber, pepper, and carrot.

The lettuces in this basic salad garden plan are all leafy, loosehead types. Traditional iceberg lettuce is difficult to grow without just the right climate. As alternate selections, the two Batavian crisphead varieties 'Sierra' and 'Nevada' offer iceberg crispness in easier-to-grow, more heat-tolerant plants. You can harvest loosehead lettuce by the leaf, beginning by taking one leaf from each plant once they have several, or you can wait and harvest entire full-grown heads. Or harvest leaves for a while, and then take the whole head. The lettuces in this plan can be replanted as long as your weather is relatively cool. Where seasons are long enough, they can also be planted again as the weather begins to cool in late summer or early fall. Some kinds tolerate high temperatures better than others, but all lettuce will grow poorly and go to seed prematurely if temperatures remain above 85° F for long.

Salad Garden Plan

1. Tasty Green cucumber
2. Garlic
3. Bell Boy sweet pepper
4. French Breakfast radish
5. Purple Plum radish
6. Red Ace beet
7. Scarlet Nantes carrot
8. Sweet 100 tomato
9. Yellow Pear tomato
10. Scallion
11. Borage
12. Arugula
13. Sweet Genovese basil
14. Parsley
15. Purple Ruffles basil
16. Chives
17. Mizuna mustard
18. Nordic spinach
19. Red Sails leaf lettuce
20. Oakleaf lettuce
21. Gem mix marigold
22. Sugar Bon snap pea
23. Medusa radicchio
24. Plato romaine lettuce
25. Buttercrunch butterhead lettuce
26. Très Fine endive
27. Johnny-jump-up

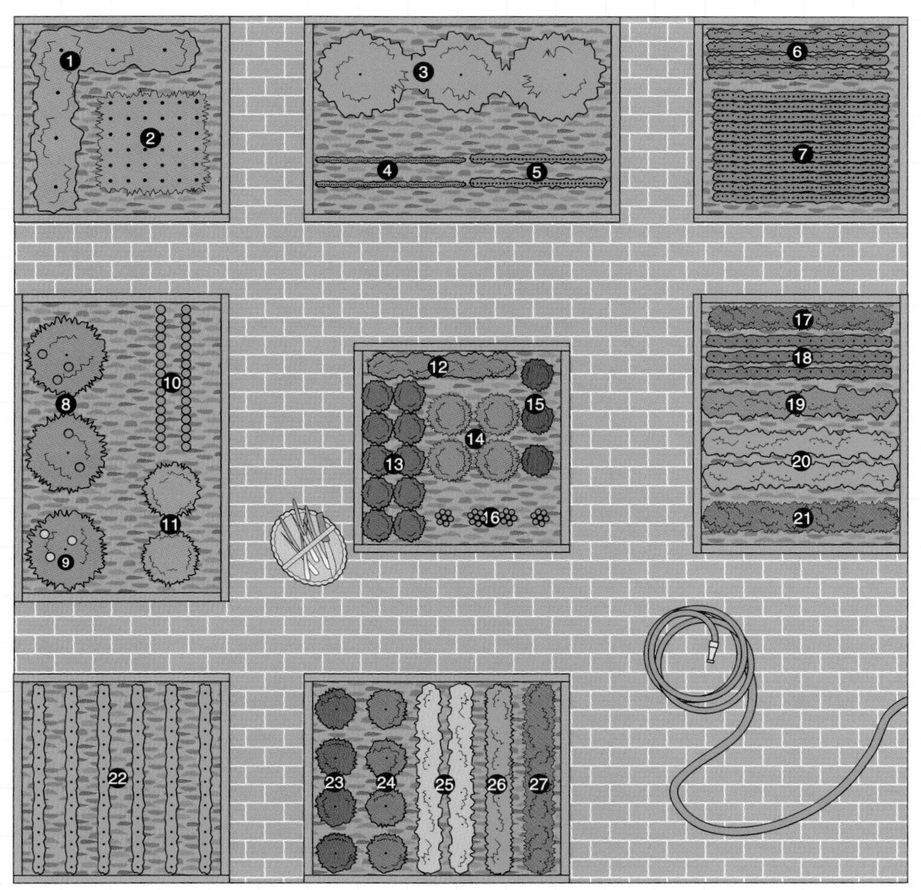

= 1 sq ft

This garden provides a mixture of mild and strong-flavored greens. The strong flavors include spicy (such as radish and arugula), bitter (radicchio and endive), and aromatic (basil and parsley). Onions, garlic, and chives provide yet another type of strong flavor. You can experiment with different combinations of greens to find the proportion of mild to strong that you and your family like best.

Small flowers, or bits of flowers, can be tossed with the salad, to make a colorful edible confetti, or whole flowers can be arranged on top. You can add a few flowers to fruit salads as well, and use them to garnish a relish plate.

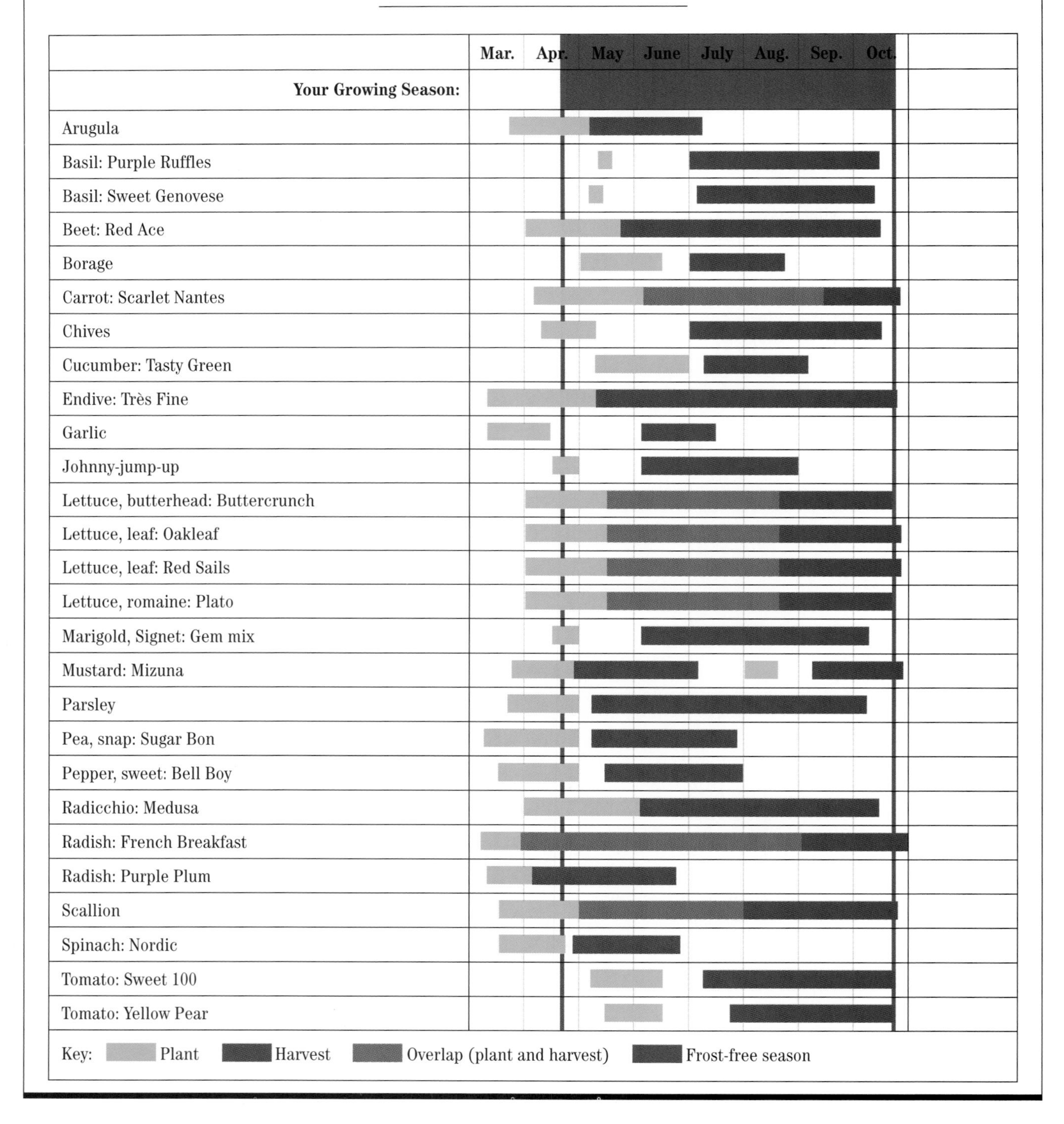

The edible flowers in this garden are those of the Gem marigold, Johnny-jump-up, borage, pea, chive, and basil. (See below for more information on edible flowers.)

Basil and garlic are choice dressing ingredients, included to give your homemade creations a fresh flavor. There's enough garlic to allow you to make a couple of braids to hang in your kitchen so you can pluck cloves as you need them.

About the Plan

This garden is designed for easy access. The beds are small and almost square, allowing you to reach in and pick greens or tomatoes quickly and easily just before dinner. Enough room is allowed for each crop to allow a succession of plantings so that you can have fresh salad makings all through the growing season. This garden would be particularly nice with the beds raised about 16 inches high and capped with a 2×6 board to serve as a seat. Because you'll be in the beds frequently, cutting fresh greens for meals, it would be pleasant to have a comfortable seat to make reaching and picking easier.

The extension to the basic plan widens the variety of salads you can make fresh from the garden, including potato salad and coleslaw. It adds 152 square feet to the basic salad garden.

Many of the vegetables in this garden have long planting and harvesting periods. They are designed for frequent small plantings, yielding a constant supply of young, fresh greens for salad. Plant small amounts of greens, scallions, and radishes every couple of weeks. Carrots and beets are planted only once, but they hold well in the soil and can be pulled all summer.

Everything in this garden can be harvested when you need it. Only the peas, beans, and cucumbers must be picked as soon as they ripen; check them regularly. If some greens get too old without being picked, harvest them and place them on the compost heap or eat them in a stew (all greens, including lettuce, can be cooked in soups and stews). Plant the potatoes in the extension plan as described on page 22, and you will be able to harvest even potatoes as you need them. New potatoes make wonderfully sweet and tender salads.

On Mesclun

Instead of buying and growing the seed of various lettuces and other salad greens separately, you can buy them premixed and sow them

On Eating Flowers

Eating flowers is not really a new idea. Many kinds of flowers have been eaten in many different cuisines. A few examples are old-time favorites such as fried squash blossoms, chamomile tea made from steeped flower heads, rose petal preserves, candied violets, and the dried daylily blossoms (golden threads) used in Chinese cooking. In the past decade there has been a revival and exploration of edible flowers in American cooking. Flowers are most often used in salads, but a glance through some recent cookbooks will show many different uses.

Some of the most popular edible flowers are included in the plant lists for this book. These include nasturtium (spicy and hot), Johnny-jump-up (mild wintergreen), borage (cucumbery), Gem marigold (citrusy), and anise hyssop (sweet and aniselike). The flowers of common cooking herbs are also edible, such as those of oregano, thyme, sage, mint, cilantro, fennel, chives, and garlic chives. In addition, many vegetables have edible flowers, including arugula, mustard, beans, peas, squash, and pumpkin. And some of our most common vegetables—cauliflower, broccoli, and artichoke—are really flower buds!

However, not all flowers are edible. Some examples of flowers that are poisonous are those of tomato, potato, eggplant, and pepper plants. Others are azalea, calla lily, daffodil, delphinium, foxglove, hyacinth, hydrangea, iris, lantana, larkspur, lobelia, lupine, oleander, poinsettia, ranunculus, rhododendron, sweet pea, and wisteria. Never eat a flower unless you are absolutely certain that the flower of the particular plant is edible and that you have identified the plant correctly.

In addition to being sure you are eating a flower that is not poisonous, you must consider whether a flower might have been treated with pesticides not registered for use on food crops. You can be most sure of this if you grow the flowers yourself or buy flowers sold in the produce section of the grocery. Do not eat flowers that you buy from a florist, because the growers may have used pesticides that are not registered for food crops.

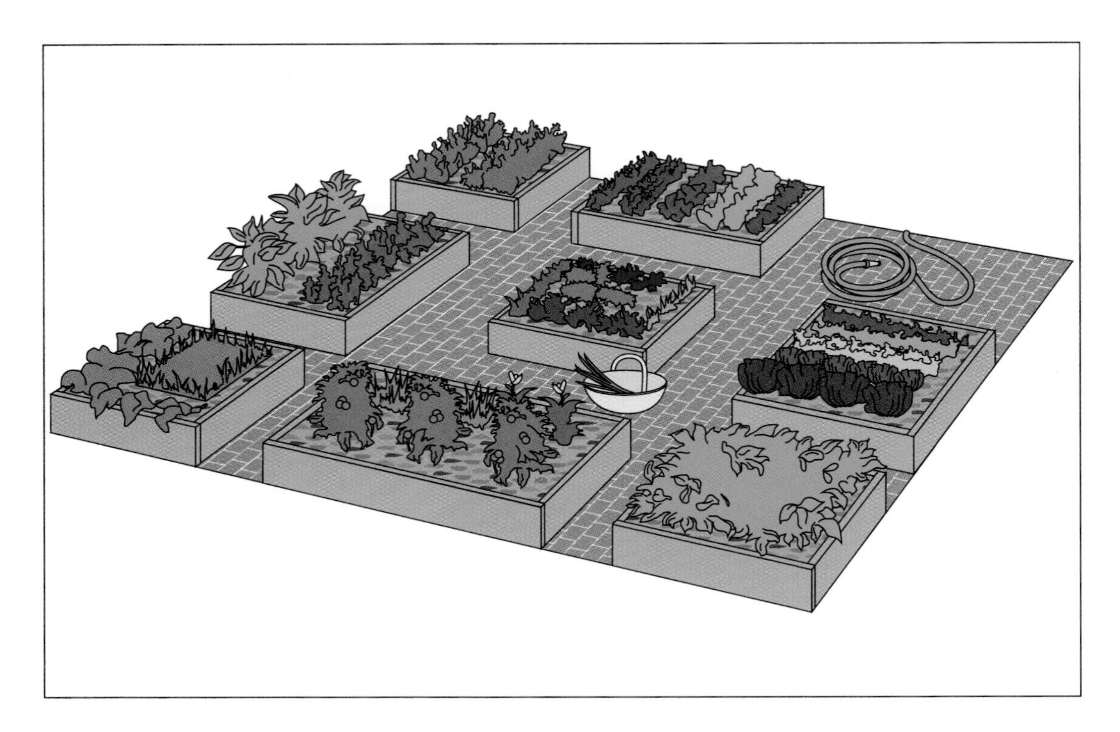

together. If the seed mixtures are scatter-sown over a garden area, they will soon form a dense block of plants. When the plants are 3 to 4 inches high, use a large pair of scissors to cut some of the plants off ½ to 1 inch from the ground. Cut a section just big enough to make one salad. When you want another salad, cut another section. The plants will regrow, giving several additional cuttings.

This idea is an old one, popular for centuries in Europe. In France, these mixtures are known as mesclun, in Italy, as misticanza or saladini. There are traditional mixes, such as Niçoise and Provençal, from the provinces of France. And now there are newer mixes, developed by American seed houses for American conditions, such as Shepherd's Napa Valley Lettuce Mix and Piquant Salad Greens Mix, or the different mild and strong mixes available from Cook's Garden.

Plant List

Arugula
Also known as rocket, roquette, and rucola. Ready in 45 days. Widely available.

Basil: Purple Ruffles
Handsome, deep purple, ruffled leaves. Flavor often less strong than green types. Ready in 60 days. Widely available.

Basil: Sweet Genovese (Genovese, Genova Profumatissima)
Or get any basil designated "large" or "large-leafed Italian." Ready in 60 days from seed. Cook's Garden, Johnny's, Nichols, Seeds of Change, Shepherd's.

Beet: Red Ace
These beets develop fast and stay tender and sweet for a long time. They have good disease tolerance. Ready in 53 days. Field's, Park, Pinetree, Territorial, Vermont Bean Seed.

Borage
Sow seed in small groups, thin to one plant per group, and add the large seed leaves of the thinnings to salads. Later, use the flowers in salads. Widely available.

Carrot: Scarlet Nantes
Cylindrical, about 6 inches long. Ready in 70 days. Widely available.

Chives
A perennial herb that is best grown using purchased plants.

Cucumber: Tasty Green 26 Hybrid
This is one of the "burpless" types with mild flavor and a thin skin. It resists powdery and

downy mildew. Ready in 62 days. Burpee, Field's, Gurney's, Nichols, Park.

Endive (frisée): Très Fine
This variety makes small heads of finely cut and curled leaves. Ready in 60 days. From Shepherd's. 'Green Curled' is similar, but the plants are twice the diameter. It is widely available.

Garlic
Grow for use in salad dressing. Purchase sets from a nursery to avoid possible diseases in grocery store garlic.

Johnny-jump-up
Easy to grow from seed. Eat whole flowers. Widely available.

Lettuce, butterhead: Buttercrunch
Forms a loose head of smooth, dark green leaves. Slow to bolt. Ready in 55 to 65 days. Widely available.

Lettuce, leaf: Oakleaf
Frilly, lobed, light green leaves have good flavor and stand up well to heat. Ready in 40 to 50 days. Widely available.

Lettuce, leaf: Red Sails
Crinkled, bronze-red leaves with good bolt resistance. Ready in 45 days. Widely available.

Lettuce, romaine: Plato
A choice romaine variety that is crisp, mild, and bolt and disease resistant. Ready in 70 days. Nichols, Pinetree.

Marigold, Signet: Gem mix
All marigolds (*Tagetes* species) are edible, but Signet marigolds (*T. tenuifolia*) have the best flavor. The Gem mix contains seeds of orange, pale yellow, and gold flowers, or you can buy single colors. Pull out petals and add to salads. Cook's Garden, Shepherd's.

Mustard: Mizuna
Harvest young, as a salad green, in spring or fall. Cut the leaves to 1 inch from the ground and the plant will regrow. Ready in 35 to 40 days. Cook's Garden, Johnny's, Pinetree.

Onion, scallion
Purchase sets in the spring and plant them out to grow scallions. Ready in 30 days. Widely available.

Parsley
The flatleaf or plainleaf types have a richer parsley flavor and are easier to chop. The curled leaf varieties are more ornamental as garnishes. It is easiest to buy plants at a local nursery.

Pea, snap: Sugar Bon
Early, sweet, and disease resistant. Use in salads either raw or lightly steamed and then chilled. Ready in 58 days. Widely available.

Pepper, sweet: Bell Boy Hybrid
A large green bell pepper that ripens to red. Resists tobacco mosaic virus. Ready in 62 to 70 days. Widely available.

Radicchio: Medusa Hybrid
One of the most reliable of the radicchio varieties, and one that can be planted in spring as well as in summer if your spring is reasonably cool. Ready in 65 days. Johnny's.

Radish: French Breakfast
These radishes are somewhat elongated, red at the top, white at the tip. Ready in 25 days. Widely available.

Radish: Purple Plum
A round radish that is a lovely purple shade, white inside. Holds its mild, sweet flavor well past maturity. Ready in 24 days. Field's.

Spinach: Nordic Hybrid
A mild-flavored, smooth-leafed type that tolerates both cool and warm weather and resists downy mildew. Ready in 39 days. Shepherd's.

Tomato, small: Sweet 100 or Supersweet 100
A heavy producer of extra-sweet, red, cherry-sized fruits. Stake or cage these large plants. Ready in 65 days. Widely available.

Tomato, small: Yellow Pear
Cherry-sized fruit, but yellow and pear shaped. Mellow flavor. Stake or cage plants. Ready in 70 days. Widely available.

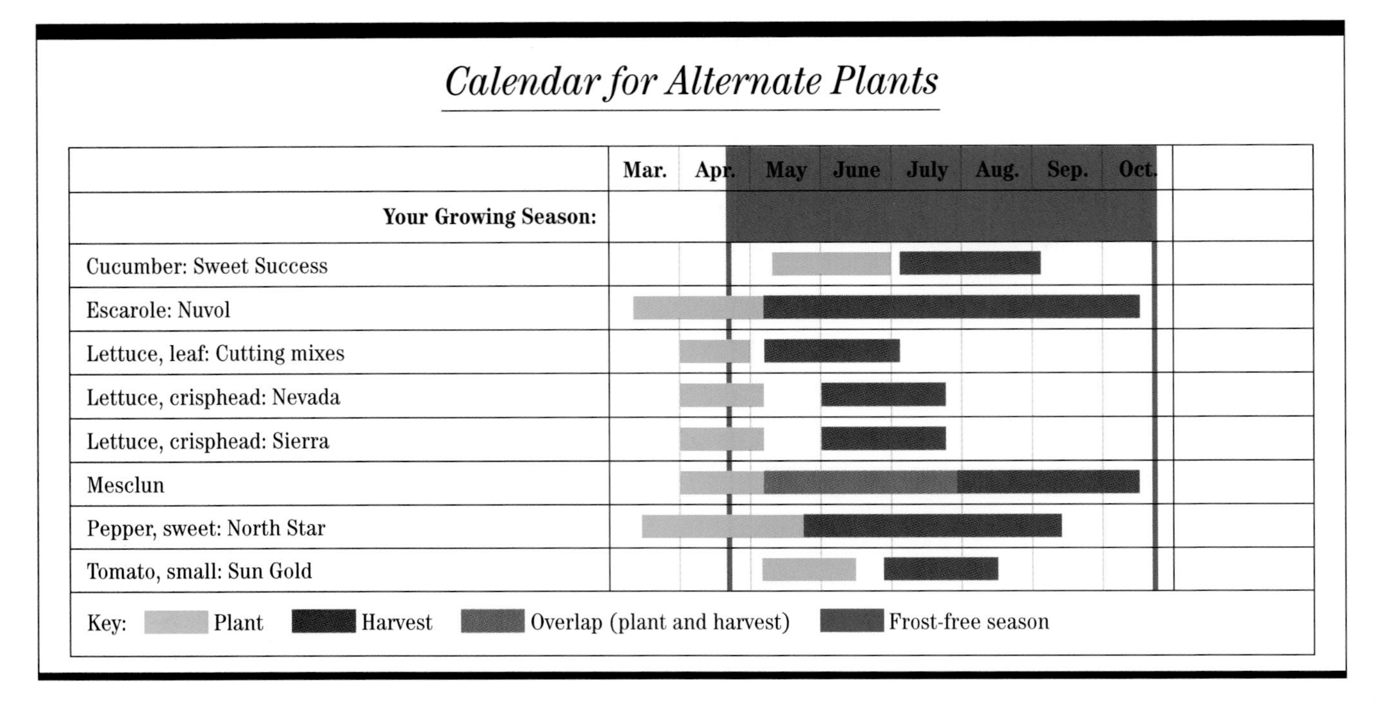

Alternate Plants

Cucumber: Sweet Success Hybrid
Try this burpless cucumber variety if you need resistance to cucumber mosaic virus, scab, or target leaf spot. Ready in 55 to 65 days. Widely available.

Escarole: Nuvol
Escarole has the same slightly bitter flavor as the curlier endive (such as 'Très Fine'), and is grown in the same way. 'Nuvol' has a self-blanched, creamy white heart and is heat resistant. Ready in 55 days. Johnny's.

Lettuce, crisphead: Nevada
A semiheading Batavian crisphead with bright green leaves. Tastes very good, resists heat very well, and can be harvested by the leaf when young, or by the whole head when mature. Ready in 60 days. Pinetree, Shepherd's.

Lettuce, crisphead: Sierra
A semiheading Batavian crisphead. You can harvest the outer leaves when the plants are young, then leave them to mature. Leaves are bright green, tinged with red. It holds up very well to heat and has a very good flavor. Ready in 70 days. Nichols, Pinetree, Territorial.

Lettuce, leaf: Cutting mixes
These carefully chosen mixes of several lettuce varieties are meant to be scatter-sown and harvested as needed by cutting sections ½ to 1 inch from the ground. Ready in 35 to 45 days. Cook's Garden, Nichols, Park, Shepherd's.

Mesclun
Various blends of strong-flavored salad greens are sold under this name. They are meant to be scatter-sown and harvested as needed by cutting sections ½ to 1 inch from the ground. Ready in 35 to 45 days. Cook's Garden, Nichols, Shepherd's, Territorial.

Pepper, sweet: North Star Hybrid
If you live in a northern, short-summer area, try this early bell pepper variety. Ready in 66 days. Widely available.

Tomato, small: Sun Gold Hybrid
This golden cherry-sized tomato has a delicious tropical flavor. Stake or cage these large plants. Ready in 57 days. Johnny's.

Plant List for Extension

Bean, pole, snap: Kentucky Wonder
Popular since the late 1800s and rust resistant. Ready in 60 to 70 days. Pole beans bear over a longer period than bush beans. Widely available.

Bean, pole, snap: Kentucky Wonder Wax
Similar to Kentucky Wonder, but the pods are pale yellow. Ready in 60 to 70 days. Gurney's.

Cabbage: Salarite Hybrid
Slightly savoyed (ruffled) leaves around a 3- to 5-pound head with a good flavor and texture for slaw. Ready in 50 to 60 days. Nichols.

Dill: Dukat or Fernleaf
Both produce plenty of leaves for use in salads. Dukat grows 3 to 6 feet tall, Fernleaf to only 18 inches. Sow them in clumps rather than as single plants. Widely available.

Kohlrabi: Early White Vienna
When peeled, the bulblike stem is creamy white. Add slices to salad or serve on the relish plate. Ready in 55 to 60 days. Widely available.

Lettuce, leaf: Red Oakleaf (Red Salad Bowl)
This lettuce has elegant lobed and frilled, deep maroon leaves. Ready in 50 days. Cook's Garden, Johnny's, Seeds of Change, Shepherd's.

Mint: Spearmint
A perennial herb that spreads by runners. You may want to grow this plant in a large, sunken pot, to control its spread. Buy a plant locally.

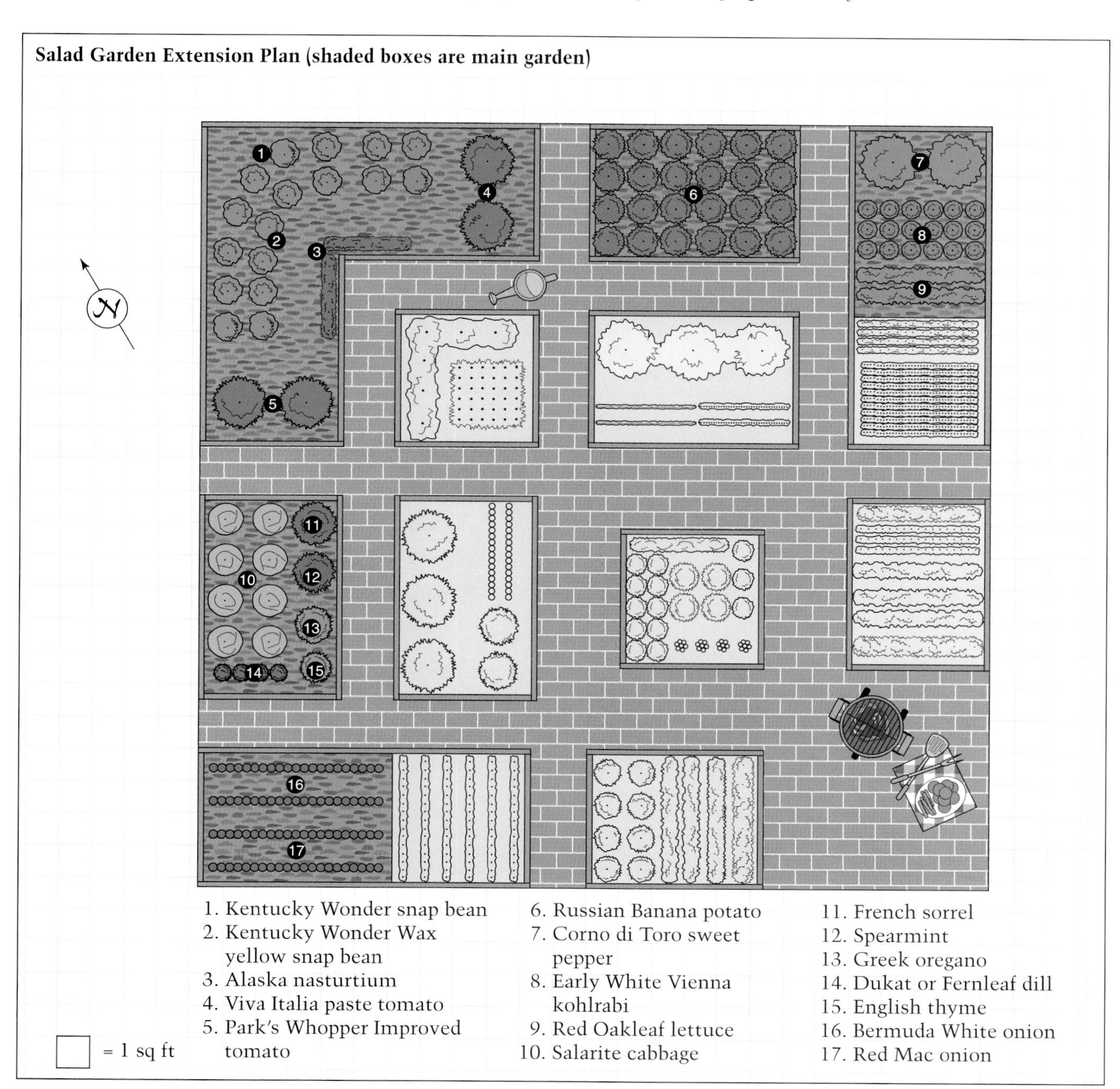

Nasturtium: Alaska (Tip Top Alaska)
Choice nasturtium selection with green-and-white variegated leaves and flowers in shades of yellow and orange. The edible leaves and flowers have a spicy flavor. Ready in 60 days. Burpee, Cook's Garden, Nichols, Pinetree, Shepherd's.

Onion plants: Bermuda White
Mild-flavored, early storage onions. Ready in about 100 days. Burpee, Park.

Onion plants: Red Mac
Large, mild, deep red onions make red-and-white rings for salads. Ready in about 100 days. Park.

Oregano, Greek
Grow this popular perennial herb to flavor salad dressings. Buy a plant locally, but rub the surface of a leaf and smell it to see if it has a good scent before you buy it.

Pepper: Corno di Toro, mixed
A 6- to 8-inch-long, sweet Italian type, good in salads or fried. Mixture of red- and yellow-ripening. Ready in 70 days. Cook's Garden, Seeds of Change.

Potato, new: Russian Banana
A small finger type with buff yellow skin and light yellow flesh, for eating as a "new" potato. Great for potato salad. Plant 18 inches apart for maximum tuber size. Mature in 90 days; "new" potatoes ready in 60 days. Johnny's.

Sorrel: French
This perennial herb adds tart flavor to salads. For just one plant, you will do best to buy a start from a local nursery.

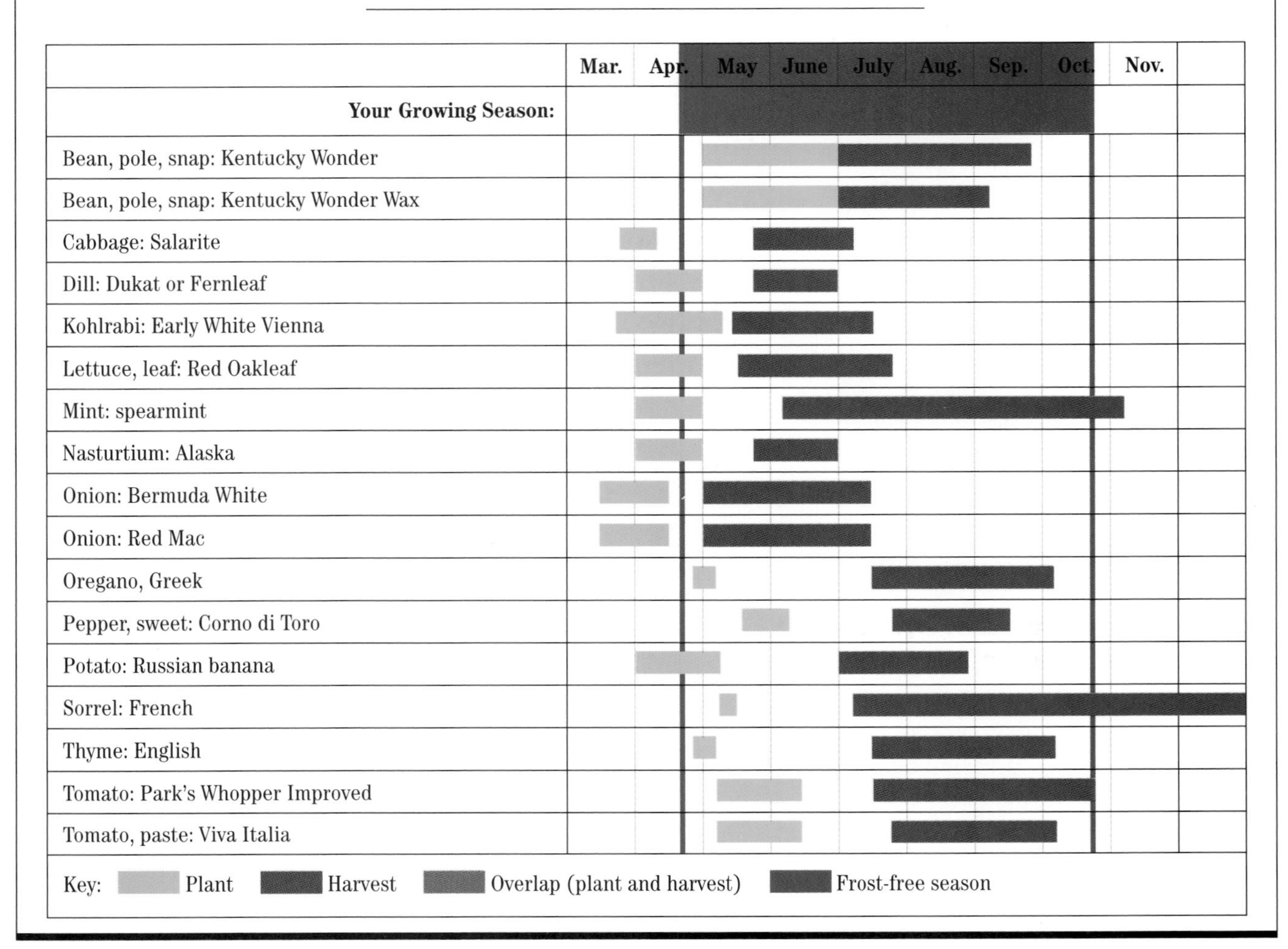

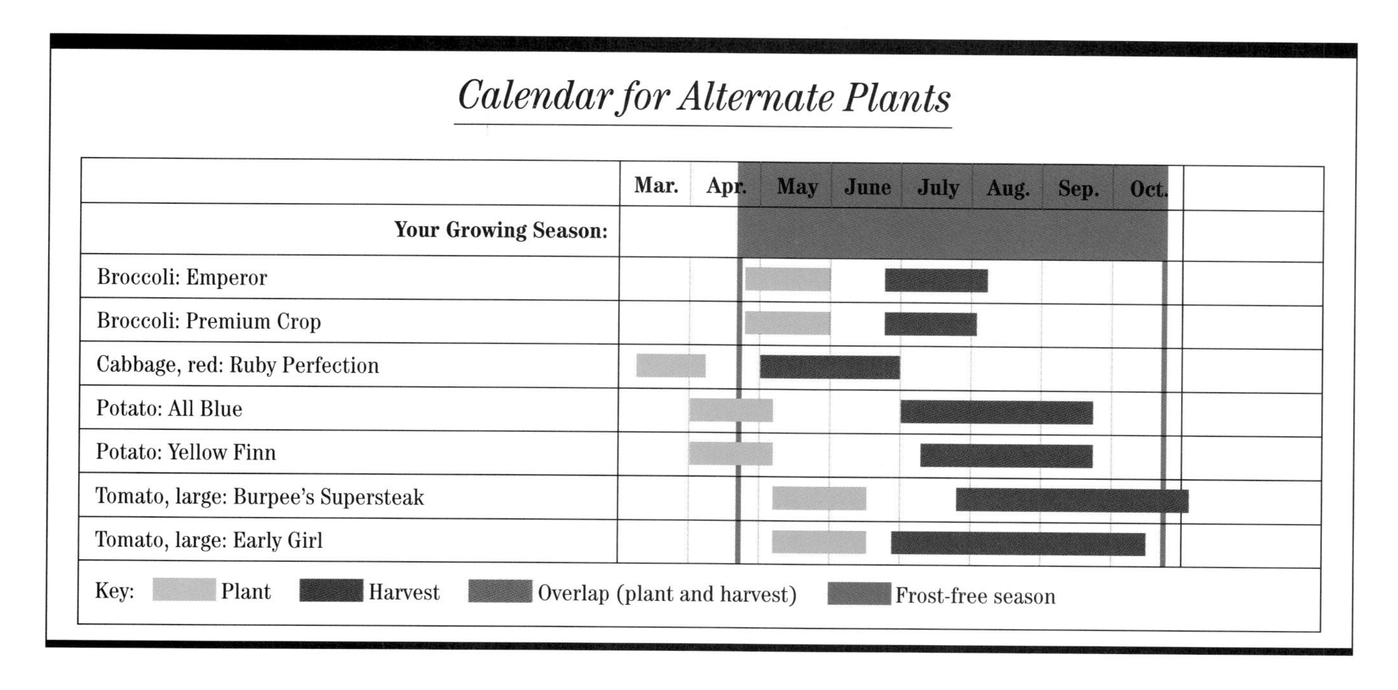

Thyme: English
A small perennial herb for seasoning salad dressings. Buy a plant locally.

Tomato, large: Park's Whopper Improved VFFNT Hybrid
Fruit is red, sweet, and juicy, up to 4 inches across. The tall plants are best staked or caged. Ready in 65 days. Park.

Tomato, paste: Viva Italia VFFNA Hybrid
Can be used for cooking, but is sweet and firm for fresh eating. It resists a number of diseases, including bacterial speck. Needs little or no staking. Ready in 80 days. Widely available.

Alternate Plants for Extension

Broccoli: Emperor Hybrid
A good spring or fall producer of large central heads and many side shoots. Tolerates black rot, downy mildew, and hollow stem. Ready in 58 to 64 days. Johnny's.

Broccoli: Premium Crop Hybrid
Makes good heads later into the summer heat than most broccolis. Produces few side sprouts, though it may make some when maturing into fall. Ready in 62 days. Widely available.

Cabbage, red: Ruby Perfection Hybrid
You may want to replace some of the green cabbages with red ones, such as this. The 3- to 5-pound heads are good for salads. Very dependable. Ready in 80 days. Cook's Garden, Johnny's.

Potato, new: All Blue
Surprising deep purple skin and flesh keep their color after cooking, to make an unusual potato salad. Ready in 110 days. Field's, Gurney's, Territorial.

Potato, new: Yellow Finn
These irregular to oval potatoes have a deep yellow skin, yellow flesh, and a delicious "buttery" flavor. Good in salad or baked. New potatoes ready in 110 days, mature ones in a few more weeks. Territorial.

Tomato, large: Burpee's Supersteak VFN Hybrid
An improvement of the traditional beefsteak tomato that offers disease resistance as well as a smaller core and blossom scar than the original. Rich-flavored fruit averages 1 pound, but some will reach 2 pounds. Ready in 80 days. Burpee, Tomato Growers Supply.

Tomato, large: Early Girl VFF Hybrid
Where summers are cool and/or short, Early Girl is a dependably early, red-fruited variety. The fruit is not huge, averaging 4 to 6 ounces each, but the flavor is very good. Ready in 52 to 62 days. Widely available.

AN ITALIAN COOKING GARDEN

Many of the plants grown in an Italian garden are familiar ones, such as beans, peppers, and lettuce, but the varieties are different. Some of them—for example, flat padded 'Romano' green beans or curled 'Annelino' beans—are seldom found in American markets.

In addition to Italian varieties of familiar vegetables, Italian cuisine includes a number of crops that Americans rarely see. Florence fennel is grown for its crisp, white leaf bases. Their flavor is aniselike when raw, becoming milder and sweeter when cooked. They can be used raw in salad, or they can be baked, grilled, or fried. The tender stems and flower buds of broccoli raab are delicious steamed and then sautéed lightly in garlic-seasoned olive oil.

Where unique Italian varieties of vegetables or herbs make a difference and are readily available, they have been suggested for use in this garden. Where standard ones will do just as well, these have been used. In some cases, standard varieties have been suggested for positive traits such as disease resistance.

About the Plan

This garden, designed to occupy a patio at the back of a house, offers a sampling of some of Italy's most famous flavors, including plenty of basil for pesto and hearty tomatoes for pasta sauce. 'Romano' beans and 'Tromboncino' squash climb trellises on either side of the garden. Herbs and salad greens near the back door let you put together a fresh salad at a moment's notice.

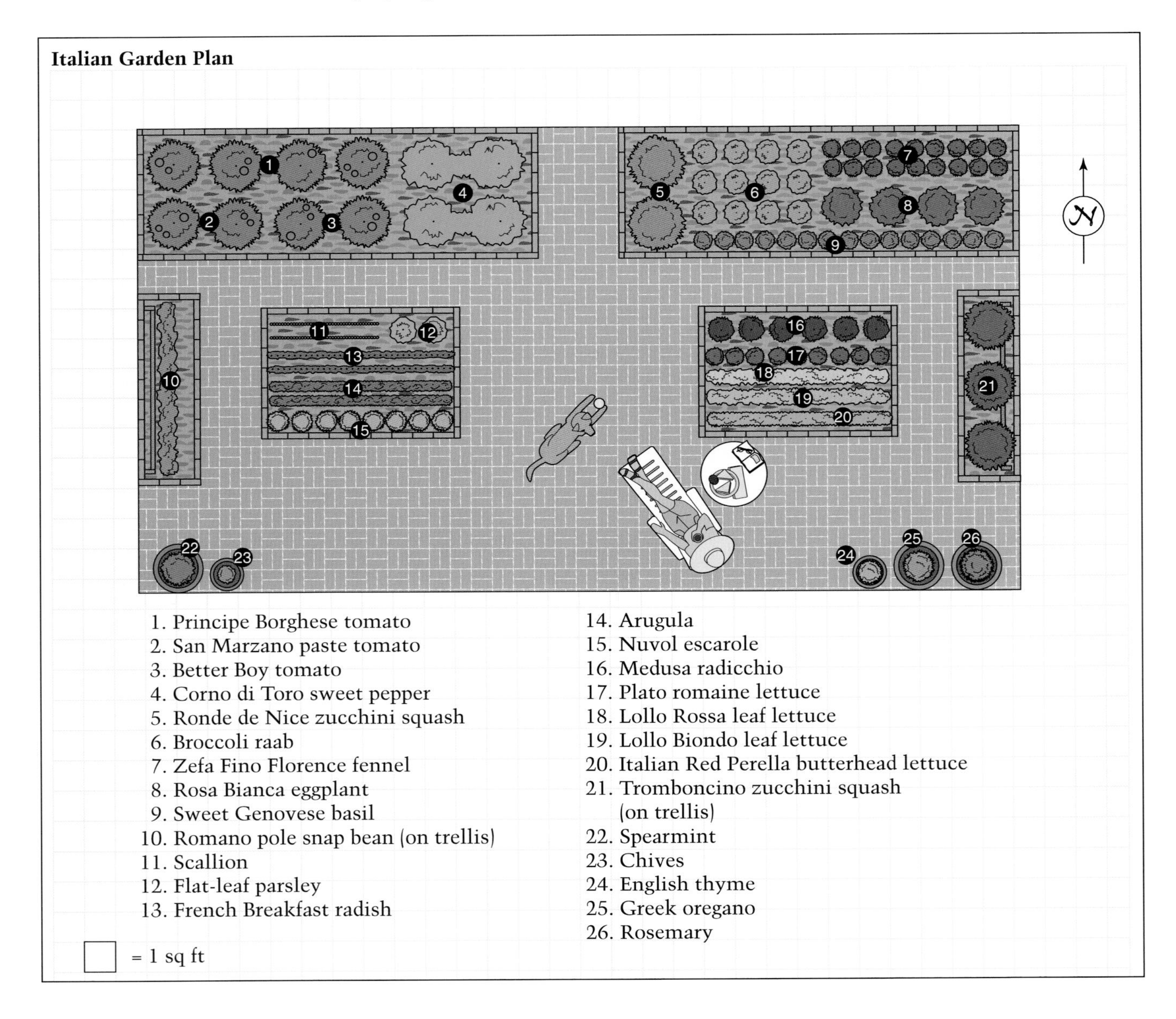

Italian Garden Plan

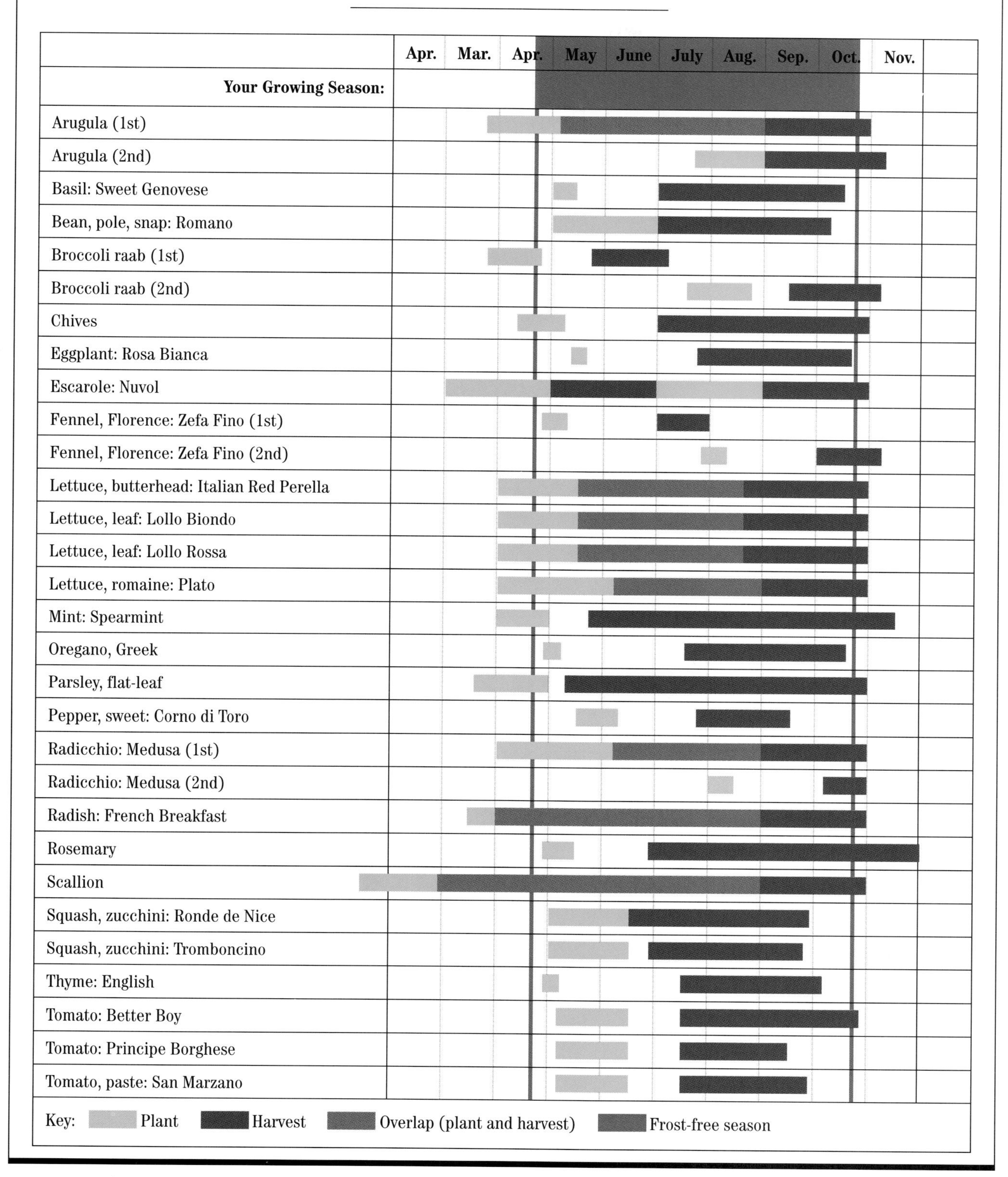

Italian Garden Calendar
Apr.
Mar.
Apr.
May
June
July
Aug.
Sep.
Oct.
Nov.
Your Growing Season:
Arugula (1st)
Arugula (2nd)
Basil: Sweet Genovese
Bean, pole, snap: Romano
Broccoli raab (1st)
Broccoli raab (2nd)
Chives
Eggplant: Rosa Bianca
Escarole: Nuvol
Fennel, Florence: Zefa Fino (1st)
Fennel, Florence: Zefa Fino (2nd)
Lettuce, butterhead: Italian Red Perella
Lettuce, leaf: Lollo Biondo
Lettuce, leaf: Lollo Rossa
Lettuce, romaine: Plato
Mint: Spearmint
Oregano, Greek
Parsley, flat-leaf
Pepper, sweet: Corno di Toro
Radicchio: Medusa (1st)
Radicchio: Medusa (2nd)
Radish: French Breakfast
Rosemary
Scallion
Squash, zucchini: Ronde de Nice
Squash, zucchini: Tromboncino
Thyme: English
Tomato: Better Boy
Tomato: Principe Borghese
Tomato, paste: San Marzano
Key:
Plant
Harvest
Overlap (plant and harvest)
Frost-free season

If you already have a patio about the right size, you can either dig holes in it or build raised beds right on top of it. A box filled with 2 feet of planting mix will grow fine vegetables, even on a concrete patio.

If you don't have a patio, you might want to put one in when you create this garden. Fill the area between the beds with crushed rock or gravel, or pave it with bricks or concrete pavers.

For those who want a larger Italian cooking garden, the extension plan nearly doubles the planting space, as well as providing a wider range of Italian vegetables. This is a good garden for a sloping yard, in the tradition of northern Italian gardens. Having a garden on an upward slope allows you an uninterrupted view of all of the plants in the garden, like the elegant 'Tromboncino' squash hanging from the trellised plants in the back bed.

If you build this plan as a hillside garden, make each bed and the path behind it the same level. Below each bed, build a retaining wall. If your property is so steep that the retaining walls would have to be more than 3 feet high, change the plan; such walls are too high. Make the beds narrower, or split them into two beds at different levels. Build a level landing where the stairs cross each path.

Plant List

Arugula
Also known as rocket, roquette, and rucola. Plant in spring and again in midsummer. Ready in 45 days. Widely available.

Basil: Sweet Genovese (Genovese, Genova Profumatissima)
An intensely flavored, perfumed basil. Use the fresh leaves to season salads or cooked dishes. Ready in 60 days. Cook's Garden, Johnny's, Nichols, Seeds of Change, Shepherd's.

Bean, pole, snap: Romano (Italian Pole)
Trellis this bean for heavy crops of broad, flat, Italian green beans. They are good for freezing. Ready in 60 days. Abundant Life, Burpee, Fields, Nichols, Seeds of Change, Vermont Bean Seed.

Broccoli raab (rapini)
This pungent green is for eating lightly steamed, rather than raw. Plant in early spring and again in midsummer. Where winters are mild, plant in the fall for an early spring harvest. Ready in 40 days, longer if overwintering. Cook's Garden, Nichols, Pinetree, Shepherd's.

Chives
A perennial herb that is best grown using purchased plants.

Eggplant: Rosa Bianca
An Italian heirloom variety with small, pink fruit. Ready in 75 days. Shepherd's.

Escarole: Nuvol
Escarole has the same slightly bitter flavor as the curlier endive and is grown in the same way. 'Nuvol' is heat resistant, with a self-

blanched, creamy white heart. Ready in 55 days. Johnny's.

Fennel, Florence: Zefa Fino
Bolt resistant for spring planting. Plant again after harvest. Where winters are mild, plant in the fall for an early spring harvest. Ready in 80 days, longer if overwintered. Cook's Garden, Johnny's, Nichols, Seeds of Change.

Lettuce, butterhead: Italian Red Perella
Three-inch leaves are green at the base and cranberry red at the tips. Flavor and texture are excellent. Grow in spring and fall when the weather is cool. Ready in 52 days. Shepherd's.

Lettuce, leaf: Lollo Biondo
Similar to Lollo Rossa, but pale green. Heat tolerant. Ready in 50 days. Cook's Garden.

Lettuce, leaf: Lollo Rossa
Compact, heat-tolerant plants with frilly leaves that are pale green in the center, tinged ruby red at the tips. Ready in 50 days. Cook's Garden, Shepherd's, Vermont Bean Seed.

Lettuce, romaine: Plato
A choice romaine variety that is crisp, mild, and bolt and disease resistant. Ready in 70 days. Nichols, Pinetree.

Mint: Spearmint
A perennial herb that needs ample water to thrive. Can be invasive. Rather than starting from seed, buy a plant.

Onion, scallion
Purchase sets in the spring and plant them out to grow scallions. Ready in 30 days. Widely available.

Oregano, Greek
A perennial herb best grown from a plant purchased locally.

Parsley: Italian flat-leaf
More flavorful than curled-leaf parsley and easier to chop. Look for various varieties of Italian origin, or just get generic flat-leaf parsley. Ready in 85 days from seed, or about 60 days from small plants. Widely available.

Pepper, sweet: Corno di Toro
A 6- to 8-inch-long, sweet Italian type, good in salads or fried. Both red- and yellow-ripening types are available. Ready in 70 days. Cook's Garden, Pinetree.

Radicchio: Medusa Hybrid
One of the most reliable of radicchio varieties, and one that can be planted in spring as well as in summer. Ready in 65 days. Johnny's.

Radish: French Breakfast
Despite the name, these radishes are as typical of Italy as they are of France. They are somewhat elongated, red at the top, white at the tip. Ready in 25 days. Widely available.

Rosemary
A perennial herb that is hardy to zone 7. Buy a plant locally. Where it is not hardy, plant it in a 6-inch pot and sink the pot in the garden for the summer; bring it indoors in winter. Widely available.

Squash, zucchini: Ronde de Nice (Round French)
Round zucchinis like this one are popular in Italy. Harvest these at 1 to 5 inches in diameter. Ready in 45 days. Cook's Garden, Seeds of Change, Shepherd's, Vermont Bean Seed. Burpee's 'Roly Poly' is similar.

Squash, zucchini: Tromboncino (Zucchetta Rampicante)
Plant runs to 5 feet or more, and may be trellised. Pick elongated, light green fruits at 10 to 15 inches. They are firm, with a mild, delicious flavor. Ready in 58 days. Cook's Garden, Pinetree, Shepherd's.

Thyme: English
A small perennial herb for seasoning salad dressings. Buy a plant locally.

Tomato, large: Better Boy VFN Hybrid
A vigorous, widely adapted tomato with sweet, juicy, red tomatoes that often weigh more than 1 pound. Plants are disease resistant, and have broad leaves to shield the fruit from sunscald. Ready in 75 days. Field's, Gurney's, Nichols, Tomato Growers Supply, Vermont Bean Seed.

Tomato, paste: San Marzano
Great flavor and productivity make this a top choice for sauce. Look for imported Italian strains such as 'La Padino', which have even better flavor. Stake or cage these tall plants. They do not have much disease resistance. Ready in 75 days. Cook's Garden, Pinetree. Pinetree has 'La Padino'.

Tomato, small: Principe Borghese
The favorite cherry-sized tomato in Italy, 'Principe Borghese' is used for drying. The fruit hangs onto the vine firmly, so the entire plant can be hung to dry. If postharvest weather will not be hot and dry, dry the fruit in an oven with a pilot light. Ready in 75 days. Cook's Garden, Pinetree.

Alternate Plants

Basil: Broadleaf Sweet
Produces an abundance of sweet, delicately flavored leaves that are excellent for pesto. Ready in 60 days. Shepherd's.

Bean, bush, snap: Roma II
Romano bean pod shape, flavor, and texture in a bush variety. Productive, disease-resistant plants bear pods that freeze very well. Ready

Calendar for Alternate Plants

	Mar.	Apr.	May	June	July	Aug.	Sep.	Oct.	Nov.
Your Growing Season:									
Basil: Broadleaf Sweet									
Bean, bush, snap: Roma II									
Broccoli: Calabrese									
Garlic									
Pepper, sweet: Bell Boy									
Squash, summer: Kuta									
Squash, zucchini: Cocozelle									
Tomato, paste: Viva Italia									

Key: Plant · Harvest · Overlap (plant and harvest) · Frost-free season

Italian Garden Extension Plan (shaded boxes are main garden)

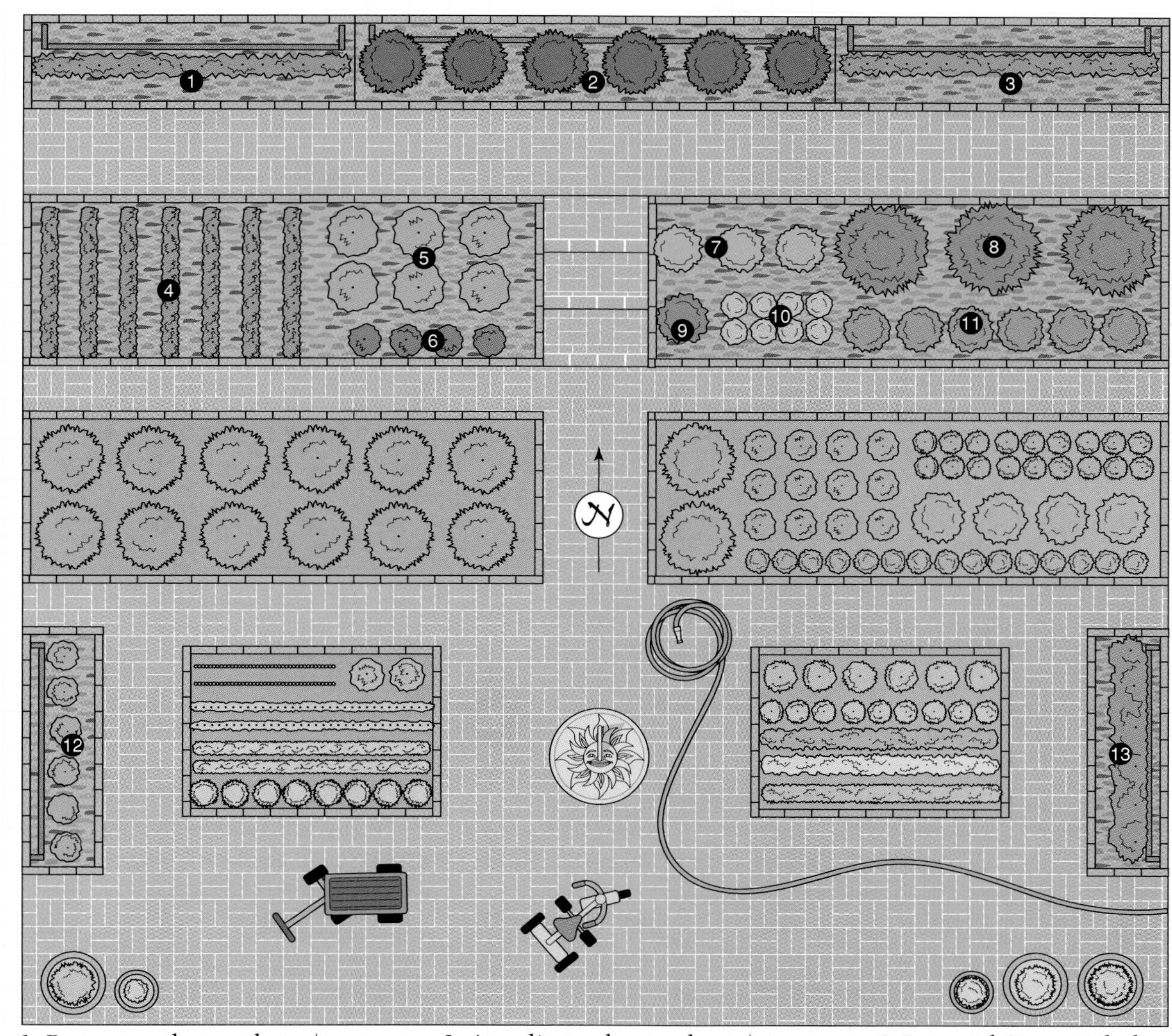

1. Romano pole snap bean (on trellis)
2. Tromboncino zucchini squash (on trellis)
3. Annelino pole snap bean (on trellis)
4. Cannellini shelling bush bean
5. Romanesco broccoli
6. Argentata Swiss chard
7. Peperoncino sweet pepper
8. Imperial Star artichoke
9. French sorrel
10. Catalogna chicory
11. Easter Egg eggplant
12. Tasty Green cucumber (on trellis)
13. Charmel melon (on trellis)

☐ = 1 sq ft

in 55 to 60 days. Burpee, Park, Shepherd's, Vermont Bean Seed.

Broccoli: Calabrese (Italian Green Sprouting)
This is the original broccoli that was introduced to American tables by Italian gardeners in 1914. The central head is 3 to 6 inches in diameter, and there are many side heads after it is cut. Ready in 58 days. Abundant Life, Pinetree, Southern Exposure.

Garlic
Purchase sets from a nursery to avoid the possibility of diseases in grocery store garlic. Widely available.

Pepper, sweet: Bell Boy Hybrid
A large green bell pepper that ripens to red. Resists tobacco mosaic virus. Ready in 62 to 70 days. Widely available.

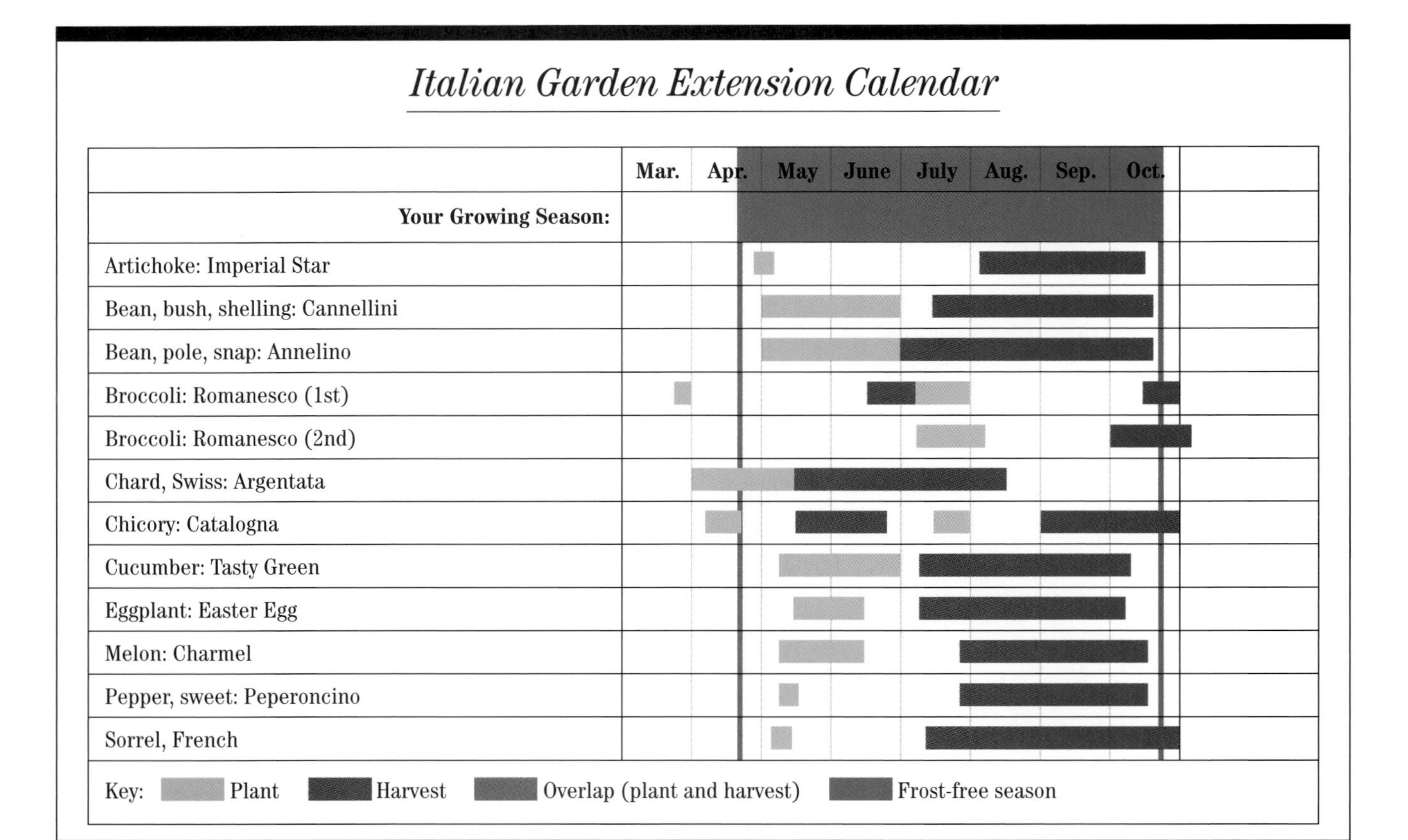

Squash, summer: Kuta Hybrid
A pale green, Middle Eastern–type summer squash that is also an Italian favorite. Ready in 48 days. Park. Similar varieties are Johnny's 'Zahra' and Cook's Garden's 'Ghada'.

Squash, zucchini: Cocozelle
An Italian type of zucchini that is striped light and dark green. There are also newer hybrids, such as 'Fiorentino', which are similar. Seeds of Change, Shepherd's. Shepherd's has 'Fiorentino'.

Tomato, paste: Viva Italia VFFNA Hybrid
Can be used in cooking, but is sweet and firm for eating fresh. Resists a number of diseases, including bacterial speck. Needs little or no staking. Ready in 80 days. Widely available.

Plant List for Extension

Artichoke: Imperial Star
Because it sets buds unusually early, this variety can be grown as an annual in cold-winter areas. Ready in 100 days. Shepherd's.

Bean, bush, shelling: Cannellini (Cannelone)
These appear in many an Italian recipe, dry or fresh-shelled. Americans often substitute dried Great Northerns, but you can try the real thing. Ready in 75 days (shelling stage). Cook's Garden, Pinetree.

Bean, pole, snap: Annelino
The pods of these heirloom Italian beans are short, flat, and coiled around into a crescent or a circle. There are green- and yellow-podded versions, both with a rich Romano-type flavor. Ready in 60 days. Cook's Garden.

Broccoli: Romanesco
The head of this unusual variety is made of chartreuse crowns arranged in a spiraling whorl. Where winters are cold, plant in mid-May. If fall and early winter are mild, plant in midsummer. Ready in 85 to 100 days. Abundant Life, Nichols, Pinetree. 'Minaret' hybrid has smaller, more uniform heads. Cook's Garden, Johnny's.

Chard, Swiss: Argentata
An heirloom Italian variety with a particularly mild, sweet flavor. It has silvery white stems and crinkled, dark green leaves. Ready in 55 days. Shepherd's.

Chicory: Catalogna (Dentarella)
Resembles our common dandelion, but is larger and more upright. Eat small in salads, or larger, cooked. Plant in early spring or mid to late summer. Ready in 40 days. Cook's Garden, Johnny's.

Cucumber: Tasty Green 26 Hybrid
This is one of the "burpless" types with mild flavor and a thin skin. Tasty Green resists powdery and downy mildew. Ready in 62 days. Burpee.

Eggplant: Easter Egg Hybrid
These look just like oversized white eggs but have a good eggplant flavor. Each 2-foot-tall plant will bear about a dozen fruits. If left on the plant, they eventually ripen to yellow, but should be eaten while still white. Ready in 52 days. Pinetree.

Melon: Charmel Hybrid
A large melon of the Charentais type, originally from France but popular throughout Europe, with smooth green skin and orange flesh. Resists fusarium and powdery mildew. Ready in 78 days. Shepherd's.

Pepper, sweet: Peperoncino (Peperoncini)
Tall, highly productive plants with small fruits. Ready in 70 to 75 days. Vermont Bean Seed.

Sorrel, French
This perennial herb adds tart flavor to salads. For just one plant, you will do best to buy a start from a local nursery.

Alternate Plants for Extension

Artichoke: Violetto
Purple artichokes top the gray-green plants of this Italian variety. Can mature in a single season if started inside in midwinter. In zones 8 to 11, start in spring and expect your first crop the following spring. Cook's Garden, Pinetree.

Cantaloupe: Ambrosia Hybrid
Extra sweet and juicy. Disease resistant. Ready in 86 days. Widely available.

Chard, Swiss: Ruby or Rhubarb
A red-stemmed chard—it is beautiful in the garden and at the table. Ready in 59 days. Widely available.

Cucumber: Bianco Lungo di Parigi
This white-skinned cucumber is very popular in Italy. It is productive and long-bearing. Ready in 65 days. Pinetree.

Calendar for Alternate Extension Plants

	Feb.	Mar.	Apr.	May	June	July	Aug.	Sep.	Oct.	Nov.	
Your Growing Season:											
Artichoke: Violetto											
Cantaloupe: Ambrosia											
Chard, Swiss: Ruby or Rhubarb											
Cucumber: Bianco Lungo di Parigi											

Key: Plant Harvest Overlap (plant and harvest) Frost-free season

AN ASIAN GOURMET GARDEN

Here is an opportunity to try a few of the many vegetables enjoyed in some of the countries of Asia. With these crops, you can prepare many common dishes from China, as well as some from Japan, Vietnam, and Thailand.

Asia has many climates, and these are reflected in its cuisines and in its gardens. Some Asian crops, such as ginger and lemongrass, are too tropical for all but the mildest of U.S. climates. And some, such as winged beans, need a long, hot, humid summer that much of the United States can't provide. But many are able to thrive here.

For this garden, preference has been given to those vegetables that are not available in most American markets. Presumably you can buy the crops that are shared by American and Asian cuisines—such as bulb and green onions, asparagus, carrots, broccoli, and spinach—or you will be growing them in another part of the garden.

Some of the crops featured in this garden are common Asian soup ingredients. Winter melon soup is a Chinese favorite. In Japan, clear soups are common, often containing just a little tofu and a few vegetables, such as mizuna, garlic chives, and snow peas. Southeast Asians make a number of "sour soups," for which the broth is seasoned with lime juice, lemongrass, and fish sauce—a salty, anchovy-based liquid. These soups are often garnished with cilantro leaves (the leaves of the coriander plant) just before they are served.

In fact, cilantro leaves are one of the most common seasonings in Southeast Asian cooking; they appear in salads, curries, and many other dishes. If cilantro is not a seasoning you have learned to like, in most dishes you can substitute mint leaves.

Another Asian plant that most Americans find to be an acquired taste is bitter melon. Many of us think that bitterness is to be avoided, but here is a crop that, due to its quinine content, is relished for it. To use this unique

Asian Garden Plan

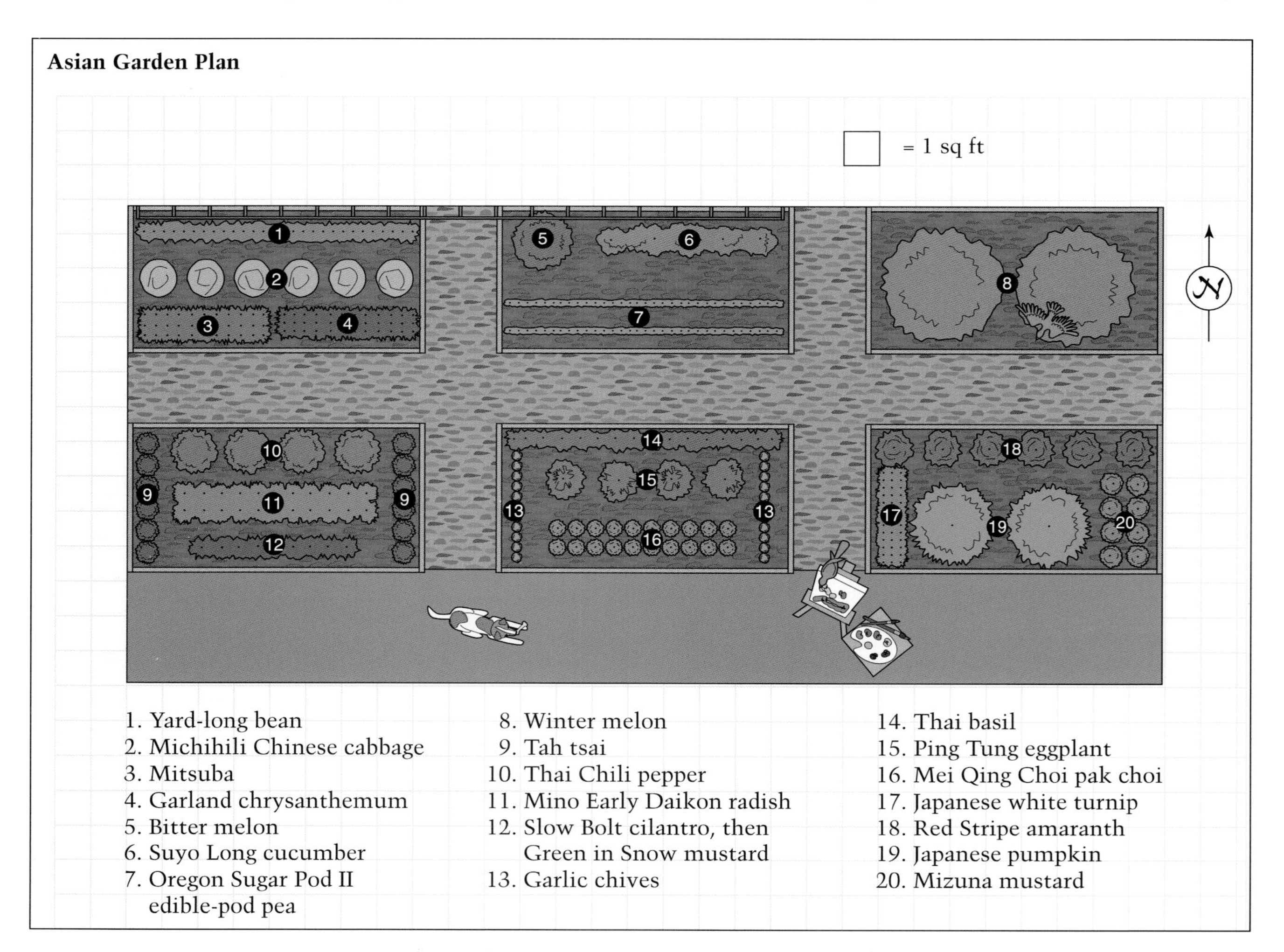

vegetable, harvest the fruits before they reach 6 inches in length, remove the (poisonous) seeds, slice the fruits, parboil them in rapidly boiling salted water for 1 to 3 minutes, then plunge them into cold water to cool them. They're now ready for use in stir-fries or soups. Bitter melon is often served with strong-flavored foods such as fish and black bean sauce.

Asian greens range from the mild Chinese cabbage (pak choi) and mizuna to sharper mustards and the distinctively flavored shungiku, which is the leaf of a chrysanthemum.

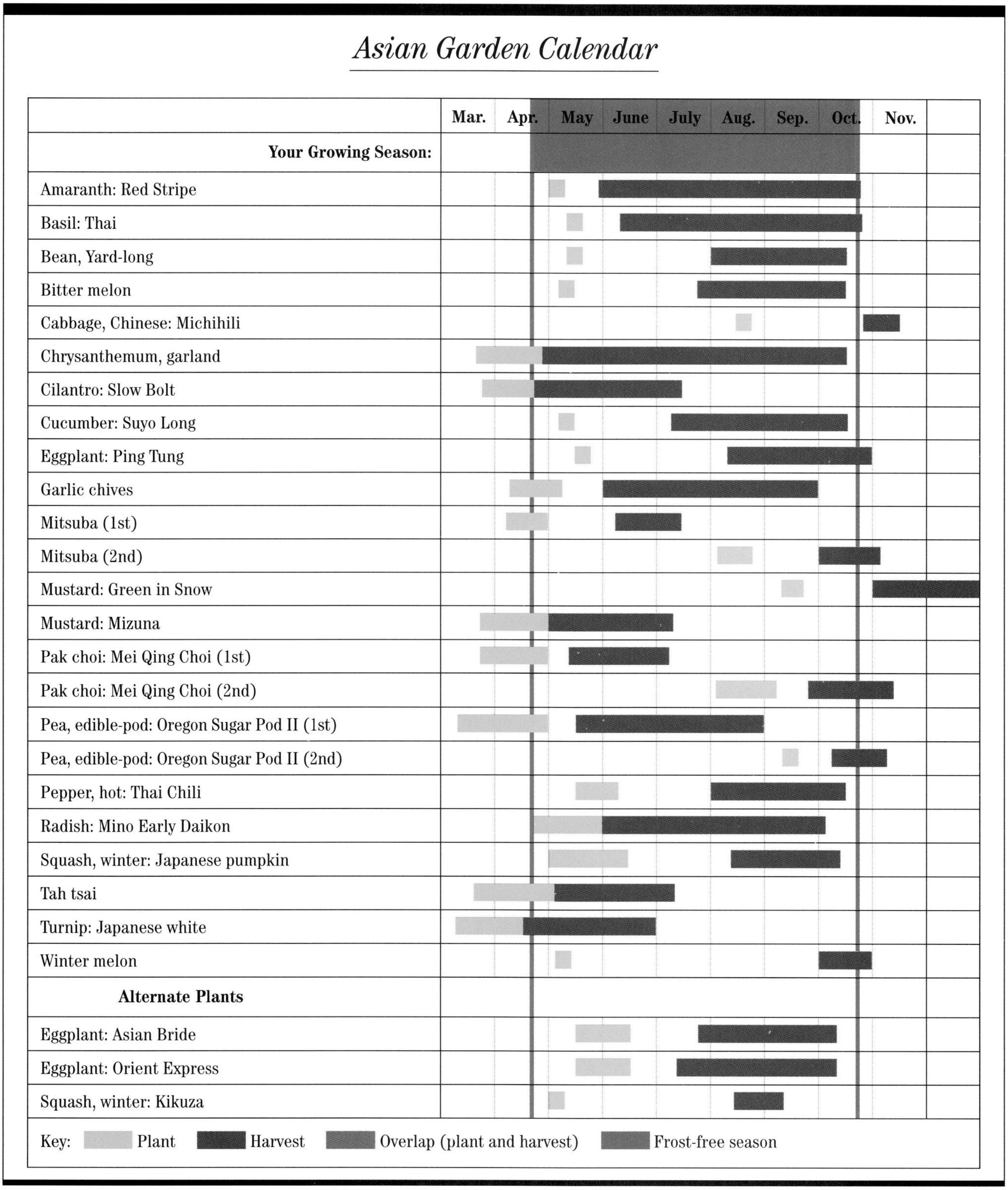

Asian Garden Calendar

	Mar.	Apr.	May	June	July	Aug.	Sep.	Oct.	Nov.	
Your Growing Season:										
Amaranth: Red Stripe										
Basil: Thai										
Bean, Yard-long										
Bitter melon										
Cabbage, Chinese: Michihili										
Chrysanthemum, garland										
Cilantro: Slow Bolt										
Cucumber: Suyo Long										
Eggplant: Ping Tung										
Garlic chives										
Mitsuba (1st)										
Mitsuba (2nd)										
Mustard: Green in Snow										
Mustard: Mizuna										
Pak choi: Mei Qing Choi (1st)										
Pak choi: Mei Qing Choi (2nd)										
Pea, edible-pod: Oregon Sugar Pod II (1st)										
Pea, edible-pod: Oregon Sugar Pod II (2nd)										
Pepper, hot: Thai Chili										
Radish: Mino Early Daikon										
Squash, winter: Japanese pumpkin										
Tah tsai										
Turnip: Japanese white										
Winter melon										
Alternate Plants										
Eggplant: Asian Bride										
Eggplant: Orient Express										
Squash, winter: Kikuza										

Key: Plant · Harvest · Overlap (plant and harvest) · Frost-free season

About the Plan

Wherever land has been scarce, intensive gardening methods have become popular, and this is true of China, where vegetables have often been grown in small raised beds, with trellises to keep sprawling vines from taking up too much room. This style of gardening was introduced to Americans in the 1980s by the Chinese-American gardener Peter Chan.

The Asian garden plan shown here includes trellises, one of which supports yard-long beans (sometimes called asparagus beans). While the pods will grow to at least a couple of feet long, they are much better when harvested at 12 to 15 inches, when they are delicious in a stir-fry with black bean sauce. Like their near relatives, southern peas, yard-longs require a long, warm growing season and will withstand hot weather. The harvest period is short, so make two plantings three weeks apart, to extend it.

If you have never tried bitter melon, try at least one plant. In short-summer areas, you will need to start this crop indoors for a few weeks. The plants need a strong trellis.

Some kinds of snow peas require a tall trellis, but the variety listed here grows no higher than 18 inches. It will thrive with no support, although it will be easier to harvest if you can provide a low fence of chicken wire or bird net on stakes. Snow peas will do best if planted either before or after summer's heat.

Garlic chives originated in southern Asia and are popular throughout the continent as well as in Japan. In China, blanched garlic chives are considered a delicacy. To produce these, harvest a mature clump by cutting the leaves an inch or so above the ground. Then cover the stumps with a large flowerpot (with the hole covered to exclude light). Harvest the blanched chives in two weeks.

Finally, the garden includes a trio of greens that are relatively new to American cooks. Tah tsai makes a low, flat rosette of dark green, spoon-shaped leaves, each with a broad, white stem. The plants hold well in warm weather. The flavor is complex and mustardy. Mitsuba is a Japanese herb that is also cooked as a vegetable. It tastes like celery and parsley, to which it is related. Mitsuba may be sown in spring or late summer and allowed to mature or used as a cut-and-come-again crop. For cut-and-come-again, broadcast seed every couple of weeks, then harvest by snipping it close to the ground with scissors. Both the narrow stems and the leaves are used in stir-fries, soups, and stews. Add mitsuba toward the end of cooking to preserve its delicate flavor. It also makes an attractive garnish. Amaranth is a warm-weather crop that grows best in the summer heat. Harvest the leafy stem tips of this delicate green to cook in any recipe that calls for spinach. In China, amaranth is often used in soups and stir-fries. Amaranth can become a weed in some areas, so remove the flowers before they go to seed.

Plant List

Amaranth (hinn choy): Red Stripe
Attractive large leaves have red stripes. Ready in 28 days. Evergreen, Sunrise.

Basil: Thai
A fragrant basil with green leaves and violet flowers. Authentic in Thai cooking, but will also make a good pesto. Ready in 60 days. Nichols, Shepherd's.

Bean, pole, snap: Yard-long (dow gauk, asparagus bean)
Bright green, slender pods, in clusters on vigorous vines to 12 feet long. Best for eating when pods are 12 to 15 inches long. Ready in 80 days. Evergreen, Gurney's, Nichols, Southern Exposure, Sunrise, Vermont Bean Seed.

Bitter melon (bitter gourd, foo gwa)
Ready in 80 days. Evergreen, Good Earth, Nichols, Kitazawa, Sunrise.

Cabbage, Chinese: Michihili (pe tsai, hakusai)
Prone to bolting if started in spring, this heading cabbage is usually grown as a fall crop, but may be tried in early spring. The flavor is sweet and delicate, and it stores well. Ready in 70 days. Widely available.

Chrysanthemum, garland (shungiku)
An edible-leaved chrysanthemum, best in cool weather. Meant to be eaten when plants are about 6 inches tall. It can be harvested by the cut-and-come-again method. Ready in 40 days. Abundant Life, Evergreen, Good Earth, Kitazawa, Sunrise.

Cilantro (coriander, Chinese parsley): Slow Bolt (Long Standing)
Ready in 30 to 40 days. Nichols, Pinetree, Shepherd's, Vermont Bean Seed.

Cucumber: Suyo Long
"Burpless," thin-skinned fruits to 14 inches long. They are deeply grooved, giving the slices an interesting shape. Disease resistant. Ready in 65 days. Evergreen, Pinetree, Sunrise.

Eggplant: Ping Tung
A choice variety of Asian-style eggplant. Ping tung fruits are tender and sweet. They reach 18 inches long, 2 inches in diameter. Evergreen.

Garlic chives (gow choy, nira)
A perennial chive that can be harvested several times a year. First cutting in 150 days. Widely available.

Mitsuba (Japanese parsley)
A perennial herb similar to parsley. First cutting in 60 days. Abundant Life, Evergreen, Good Earth, Kitazawa, Sunrise.

Mustard: Green in Snow
A very hardy, spicy green for fall or winter use. Ready in 40 days. Evergreen, Good Earth.

Mustard: Mizuna
Harvest young leaves in spring or fall. Ready in 35 days. Evergreen, Kitazawa, Pinetree, Sunrise.

Pak choi: Mei Qing Choi
A small, green-stemmed type, easy to grow in spring or fall. Ready in 45 days. Cook's Garden, Good Earth, Johnny's, Pinetree.

Pea, edible-pod: Oregon Sugar Pod II
A stringless pod on a disease-resistant plant that grows to 28 inches. Plant in spring or late summer. Ready in 68 days. Abundant Life, Evergreen, Nichols.

Pepper, hot: Thai Chili (Lo Chaio)
A heavy yielder of small, very hot peppers that ripen to red. Ready in 80 days. Evergreen, Good Earth, Pinetree.

Radish: Mino Early Daikon
A mild white radish that tolerates summer heat. To 3 inches by 30 inches. Ready in 45 days. Evergreen, Sunrise.

Squash, winter: Japanese pumpkin
You will find various strains of Japanese pumpkin or Kabocha squash, all rounded and dark green with deep orange flesh. Ready in 105 days. Evergreen, Johnny's, Kitazawa, Good Earth, Pinetree.

Tah tsai (tatsoi)
A fast and dependable crop of small, dark green leaves with broad white stems. Ready in 45 days. Abundant Life, Johnny's, Pinetree.

Turnip: Japanese white
Look for the small, white Japanese turnips meant to be harvested young for baby greens and tender, mild roots. Varieties include Hakurei, Tokyo Cross, and Tokyo Market. Ready in 38 days. Evergreen, Good Earth, Johnny's, Kitazawa, Nichols, Pinetree.

Winter melon (doan gwa, wax gourd)
A large, vigorous gourd that can be harvested immature as a "fuzzy melon" or allowed to mature into a "wax melon" that is used to make the festive winter melon soup. Ready in 90 days. Evergreen, Good Earth, Nichols, Sunrise.

Alternate Plants

Eggplant: Asian Bride
Very productive and pretty, these 2-foot-tall plants have elongated fruits that are white streaked with lavender. The light green leaves attract fewer flea beetles than do purple-tinged eggplant leaves. Ready in 70 days. Shepherd's.

Eggplant: Orient Express Hybrid
Here is an extra-early eggplant with the ability to set fruit in extremes of hot and cool weather. The fruit is slender, and 8 to 10 inches long. Ready in 58 days. Johnny's.

Squash, winter: Kikuza
A squash that originated in southern China. It is ribbed, buff-colored, and weighs 4 to 5 pounds. The flesh is orange, sweet, and fine-textured. Ready in 100 days. Nichols.

A MEXICAN COOKING GARDEN

The food served in most Mexican restaurants in the United States represents only a narrow sample of the many delicious dishes of this lively cuisine. Cookbooks show us much more of the wide array of salsas, salads, soups, vegetable and seafood dishes, and meats cooked in various sauces that are enjoyed south of the border. A Mexican cooking garden can provide you with the fresh ingredients to re-create many of these dishes in your kitchen.

Many vegetables popular in Mexico are familiar to us: Tomatoes, onions, corn, chilies, and beans are all staples of Mexican cuisine. Swiss chard, potatoes, and summer squash are

Mexican Garden Plan

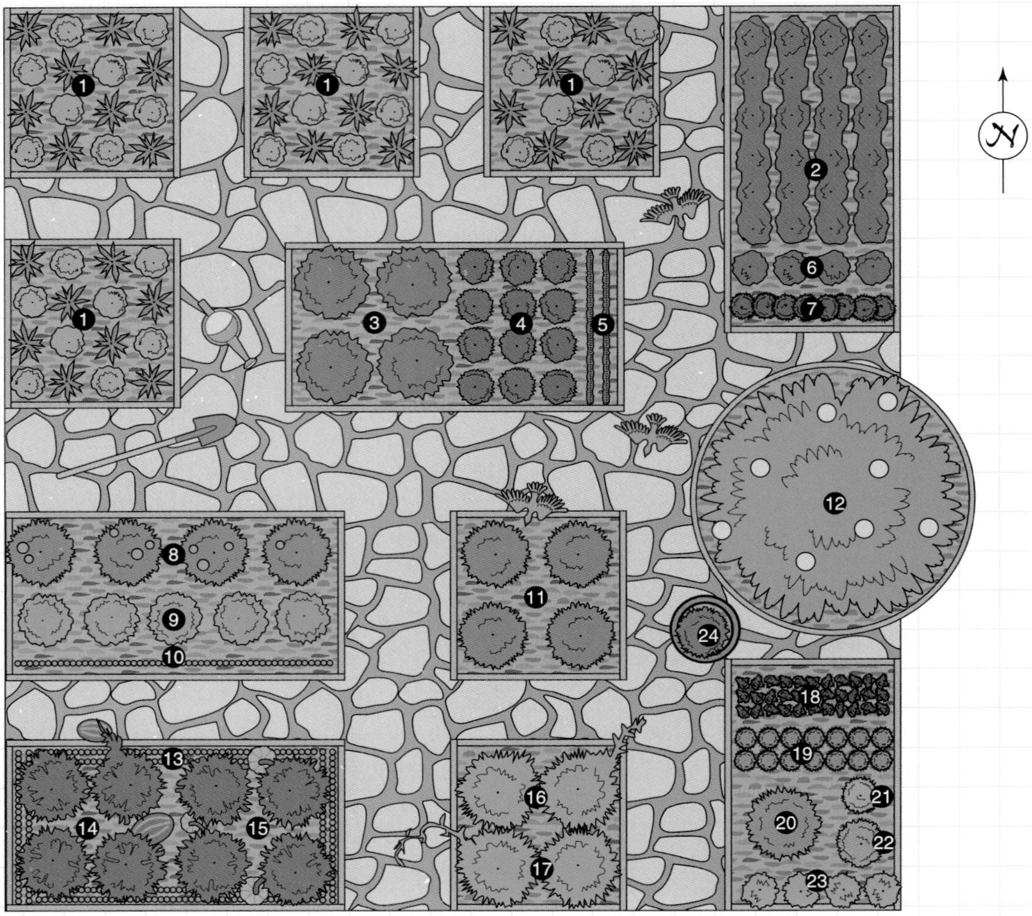

1. Kandy Korn sweet corn, interplanted with Kentucky Blue pole snap beans
2. Red Pontiac potato
3. Poblano/ancho semihot pepper
4. Plato romaine lettuce
5. Cherry Belle radish
6. Fordhook Giant Swiss chard
7. Slow Bolt cilantro
8. Viva Italia paste tomato
9. Jalapeño hot pepper
10. Scallion
11. Tomatillo
12. Small fruit tree
13. Bermuda White onion
14. Sugar Baby watermelon
15. Galia melon
16. Greyzini zucchini squash
17. Ronde de Nice zucchini squash
18. Verdolaga purslane
19. Quelite
20. Epazote
21. English thyme
22. Greek oregano
23. Flat-leaf parsley
24. Spearmint

☐ = 1 sq ft

also commonly used. Some of the vegetables grown in Mexico are less familiar—for example, *quelites*. This term can refer to any of several wild greens that are allowed to reproduce in gardens for use in cooking, but it is often used specifically for certain species of lamb's-quarters (*Chenopodium* species). While lamb's-quarters grows wild in much of the United States, and some species are eaten, to see what Mexican recipes intend, you can order some *Chenopodium berlandieri* from Native Seeds/ Search, a seed company that specializes in crops of Native Americans of the Southwest on both sides of the border.

Another wild green eaten in Mexico is purslane, known there as *verdolaga*. This plant (*Portulaca oleracea*) is a common weed in many U.S. gardens. If you have some, you can just harvest young, succulent plants before they bloom. If there is none already in your yard, you could order seed for a domesticated variety, much larger and more upright than the

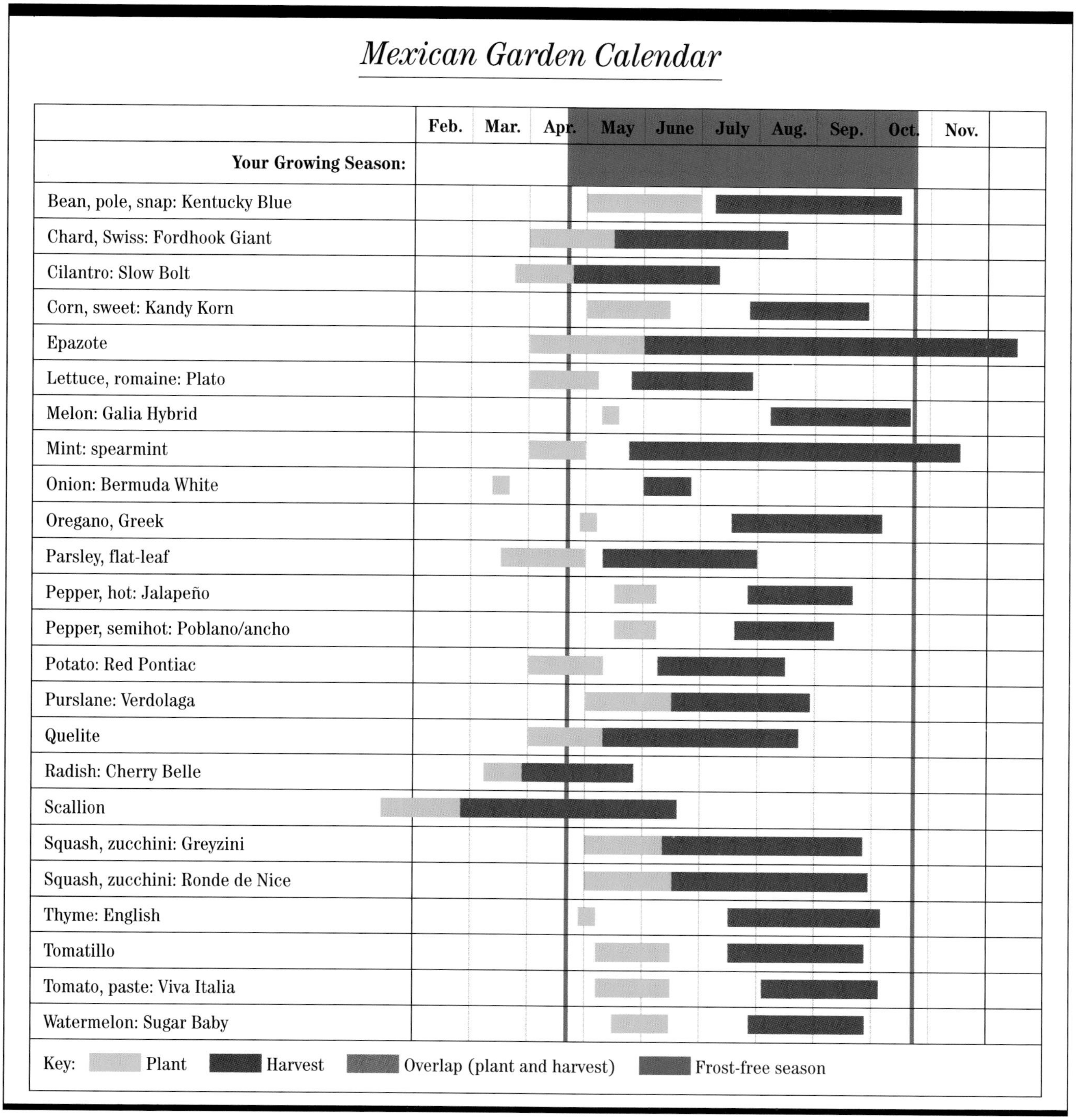

Mexican Garden Calendar

	Feb.	Mar.	Apr.	May	June	July	Aug.	Sep.	Oct.	Nov.
Your Growing Season:										
Bean, pole, snap: Kentucky Blue										
Chard, Swiss: Fordhook Giant										
Cilantro: Slow Bolt										
Corn, sweet: Kandy Korn										
Epazote										
Lettuce, romaine: Plato										
Melon: Galia Hybrid										
Mint: spearmint										
Onion: Bermuda White										
Oregano, Greek										
Parsley, flat-leaf										
Pepper, hot: Jalapeño										
Pepper, semihot: Poblano/ancho										
Potato: Red Pontiac										
Purslane: Verdolaga										
Quelite										
Radish: Cherry Belle										
Scallion										
Squash, zucchini: Greyzini										
Squash, zucchini: Ronde de Nice										
Thyme: English										
Tomatillo										
Tomato, paste: Viva Italia										
Watermelon: Sugar Baby										

Key: Plant Harvest Overlap (plant and harvest) Frost-free season weed. To try verdolaga, sauté a small onion, finely chopped, and add a couple of cups of chopped young verdolaga stems and leaves. Moisten this combination with about ½ cup tomato sauce; add a touch of finely chopped jalapeño pepper.

If you have eaten a green salsa, you have tried tomatillo. This close relative of the tomato grows covered with a papery husk, which is easily removed prior to use. It is most commonly eaten green, and is the basis for many a green salsa or green cooking sauce. Tomatillos are as easy to grow as tomatoes.

A seasoning unique to Mexican cooking is *epazote*. It is used especially in bean dishes, but also in other cooking. The plant is *Chenopodium ambrosioides*, a close relative of lamb's-quarters. It is not likely to be mistaken for lamb's-quarters, however, because its scent and flavor are very strong. A little clearly goes a long way! This plant is also sometimes seen as a weed in the United States and may self-sow profusely if it is allowed to go to seed, so keep the flowering heads cut back.

The melon in this plan is a proven, popular variety developed in the Middle East. If you want to experiment, try one of the Southwestern varieties sold by Native Seeds, such as Chimayo, an heirloom melon from northern New Mexico, or the Rio Grande Pueblo Mix.

About the Plan

Pre-Columbian Mexican gardens, called *milpas*, were casual mixes of trees, vines, vegetables, and fruits, and gardens of modern Mexico often follow this design. The plan shown on page 46 echoes some of the aspects of the ancient *milpas*. It has an asymmetrical pattern that includes a fruit tree. (The tree could be lemon or lime, if you have a climate warm enough for them. If not, try a peach or persimmon.) The herbs and greens are arranged in rows in the plan, but some, including the verdolaga, quelite, and cilantro, can be tucked into any available corner, where they will eventually reseed themselves.

In the four small back beds, corn is intertwined with pole beans. For this to succeed, take a couple of precautions. First, be sure that the corn you choose is a tall-stalked one; 'Kandy Korn' generally attains 8½ feet. Then be sure that the corn is well up—say, a foot tall—before you put in the beans. Plant two or three beans on the north side of each set of four corn stalks.

In the left front bed, plant a wide border of bulbing onions around the bed. The onions will mature shortly after the melons are planted. While the onions are drying, water only the melons in the center of the bed. After the onions are harvested, you can water the entire bed.

Plant List

Bean, pole, snap: Kentucky Blue
These are sweeter than 'Kentucky Wonder'. The plants bear an abundant crop of uniform 7-inch pods. Ready in 65 days. Pole beans bear over a longer period than bush beans. Widely available.

Chard, Swiss: Fordhook Giant
Chard is featured in a number of Mexican vegetable recipes. Ready in 60 days. Widely available.

Cilantro (coriander): Slow Bolt (Long Standing)
Ready in 30 to 40 days. Nichols, Pinetree, Shepherd's, Vermont Bean Seed.

Corn, sweet: Kandy Korn EH Hybrid
An extra-sweet yellow corn with 8½-foot stalks and 8-inch ears. Unlike other super-sweet corns, it does not require isolation from other varieties. Holds flavor well for two weeks on the stalk. Ready in 89 days. Burpee, Gurney's, Shepherd's, Vermont Bean Seed.

Epazote (*Chenopodium ambrosioides*)
A plant related to lamb's-quarters that is used in Mexico to season food, especially beans. Ready in 45 to 90 days. Abundant Life, Johnny's, Native Seeds, Nichols, Park, Pinetree, Shepherd's, Southern Exposure.

Lettuce, romaine: Plato
A choice romaine variety that is crisp, mild, and bolt and disease resistant. Ready in 70 days. Nichols, Pinetree.

Melon: Galia Hybrid
A honey-sweet, green-fleshed melon that produces well even where nights are somewhat cool. Of Middle Eastern origin, this melon is a type widely used in Mexico. Ready in 85 days. Cook's Garden, Shepherd's.

Mint: Spearmint
A perennial herb that needs ample water to thrive. Can be invasive. Rather than starting from seed, buy a plant. Grow in a large container and never let soil dry out.

Onion, bulbing: Bermuda White
Mild-flavored, early storage onions. Ready in about 100 days. Burpee, Park.

Onion, scallion
Purchase sets in the spring and plant them out to grow scallions. Ready in 30 days. Widely available.

Oregano, Greek
A perennial herb best grown from a plant purchased locally.

Parsley: Italian or Flat-leaf
The flat-leaf types have a richer parsley flavor and are easier to chop. Start from seed, or purchase seedlings locally.

Pepper, hot: Jalapeño
The regular, hot, non-early type. Grow where summer really heats up and you want a seriously hot jalapeño. Ready in 72 to 80 days. Widely available.

Pepper, semihot: Poblano/ancho
When green, this broad, 4-inch-long pepper is called poblano, and is the favored Mexican stuffing pepper (for chiles rellenos, for instance). When dried, it is a dark red, mildly pungent pepper used in cooking sauces, and is called ancho. Will bear best where summers are hot. Ready in 70 to 80 days. Widely available.

Potato: Red Pontiac
Round, red-skinned potatoes that can be dug when small, at the "new potato" stage, or left to get larger. Mature in 90 to 100 days. Widely available.

Purslane: Verdolaga (*Portulaca oleracea*)
A very large, upright form of purslane. If allowed to form seeds, it could become weedy. Ready in 60 days. Pinetree, Territorial.

Quelite (*Chenopodium berlandieri*)
Mexicans gather this plant wild or let it self-sow in their gardens to use as a green. It is a relative of our common weed, lamb's-quarters. Native Seeds.

Radish: Cherry Belle
Round and bright red. Ready in 22 days. Widely available.

Squash, summer, zucchini: Greyzini (Grey Zucchini)
Light green zucchini with faint stripes and gray mottling. This one is extremely productive, so be ready to pick them small. Ready in 42 days. Field's, Gurney's, Pinetree, Southern Exposure.

Squash, summer, zucchini: Ronde de Nice (Round French)
A pale green, round zucchini of the type favored by Mexican cooks, this one is a European heirloom. Harvest them when they are 1 to 5 inches in diameter. Ready in 45 days. Cook's Garden, Seeds of Change, Shepherd's, Vermont Bean Seed.

Thyme: English
A small perennial herb for flavoring soups and stews. Buy a plant locally.

Tomatillo or husk tomato
You'll find that tomatillos are widely available, but the fruit varies in size and flavor. 'Indian Strain', offered by Territorial Seeds, is one of the better-tasting ones. 'Purple de Milpa' has purple-tinged ⅝-inch berries. Ready in 60 to 90 days. Abundant Life, Nichols, Seeds of Change, Southern Exposure.

Tomato, paste: Viva Italia VFNA Hybrid
Paste tomatoes are best for use in Mexican-style salsas and cooking sauces. This variety is also sweet and firm for eating fresh. It resists a number of diseases, including bacterial speck. Needs little or no staking. Ready in 80 days. Widely available.

Watermelon: Sugar Baby
An old-garden favorite variety. It has round, sweet, red-fleshed fruits with few seeds, averaging 6 to 8 inches in diameter and 8 to 10 pounds. Ready in 75 days. Widely available.

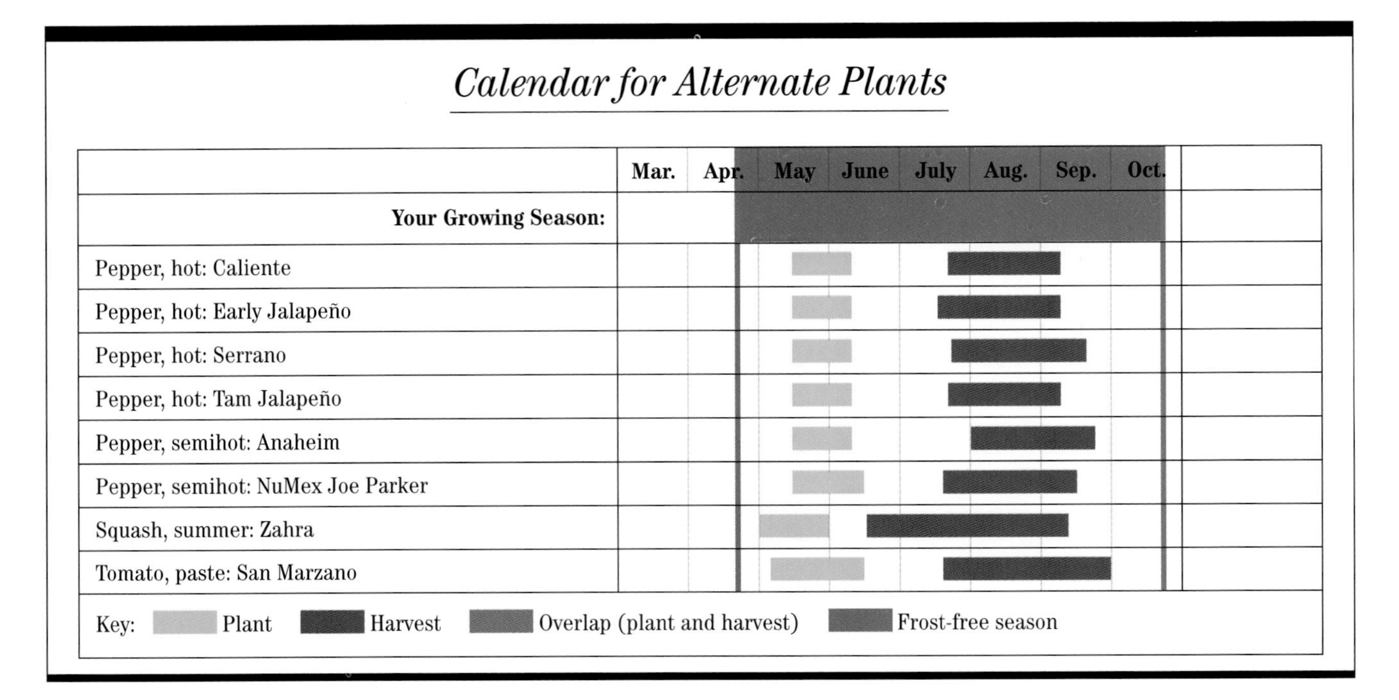

Alternate Plants

Pepper, hot: Caliente Hybrid
An early, 6-inch-long hot pepper that ripens to red. The thin walls of this pepper dry easily, making it good to string for decoration or grind into chili powder. Green-ripe in 65 days. Johnny's.

Pepper, hot: Early Jalapeño
Hot green peppers to 3 inches long, ripening to red. Use in salsas and cooking sauces and for pickling. Where conditions are cool, this variety ripens fruit better than other jalapeño strains or, for that matter, than most other hot peppers. Ready in 63 to 66 days. Abundant Life, Johnny's, Pinetree, Territorial.

Pepper, hot: Serrano
These peppers are smaller than jalapeños, only 2 inches by ½ inch, and very hot. They are widely used in Mexican recipes, especially when green, and they can also be dried. Ready in 70 days. Burpee, Field's, Gurney's, Nichols, Pinetree, Seeds of Change, Shepherd's, Southern Exposure, Tomato Growers Supply.

Pepper, hot: Tam Jalapeño
For those who prefer less fire, a jalapeño that is only a quarter as hot as the other common strains. Ready in 65 to 70 days. Field's, Gurney's, Nichols, Tomato Growers Supply.

Pepper, semihot: Anaheim
Long, tapered, with a sweet to slightly warm flavor, these can be used for stuffing. Ready in 77 days. Widely available.

Pepper, semihot: NuMex Joe Parker
A mild pepper, suitable for stuffing, that ripens two weeks before 'Anaheim'. It is 5 to 6 inches long and, while it is fairly hot, is less so with the seeds and membranes removed. Ready in 65 days. Shepherd's, Tomato Growers Supply.

Squash, summer: Zahra Hybrid
Although this squash is listed as a Middle Eastern type, Central and South Americans will recognize it as a type called pipian. Its blossoms are large and firm, and harvesting most of the male blossoms will not reduce the fruit harvest. Ready in 46 days. Johnny's.

Tomato, paste: San Marzano
Great flavor and productivity. Look for imported Italian strains such as 'La Padino', which have even better flavor. Stake or cage these tall plants. They do not have much disease resistance. Ready in 75 days. Cook's Garden, Pinetree. Pinetree has 'La Padino'.

AN HERB GARDEN

The meaning of the word *herb* has changed over time. Long ago, it meant any small, non-woody plant. Now, we use the term mainly to indicate any plant grown for use as a seasoning for cooking, for its scent, or for its medicinal properties; some would also include plants grown to create fabric dyes.

However, of most interest to a vegetable gardener are the herbs that are used to enliven our meals. This garden will let you sample many of the most common cooking herbs. You will be able to grow the plants that produced those little jars of dried leaves, and have the delicious fresh herbs to use all summer—or even longer where winters are mild.

About the Plan

The basic herb garden shown here is a kitchen herb garden; the herbs are all useful in cooking. Many are herbs that taste best when used fresh (if you are accustomed to them in their commercially dried forms, you may be surprised—and delighted—by their flavorfulness). Included are also some herbs grown mainly for their seeds or roots.

The design is a theme common in herb gardens: a circle around a cross. This is an ancient

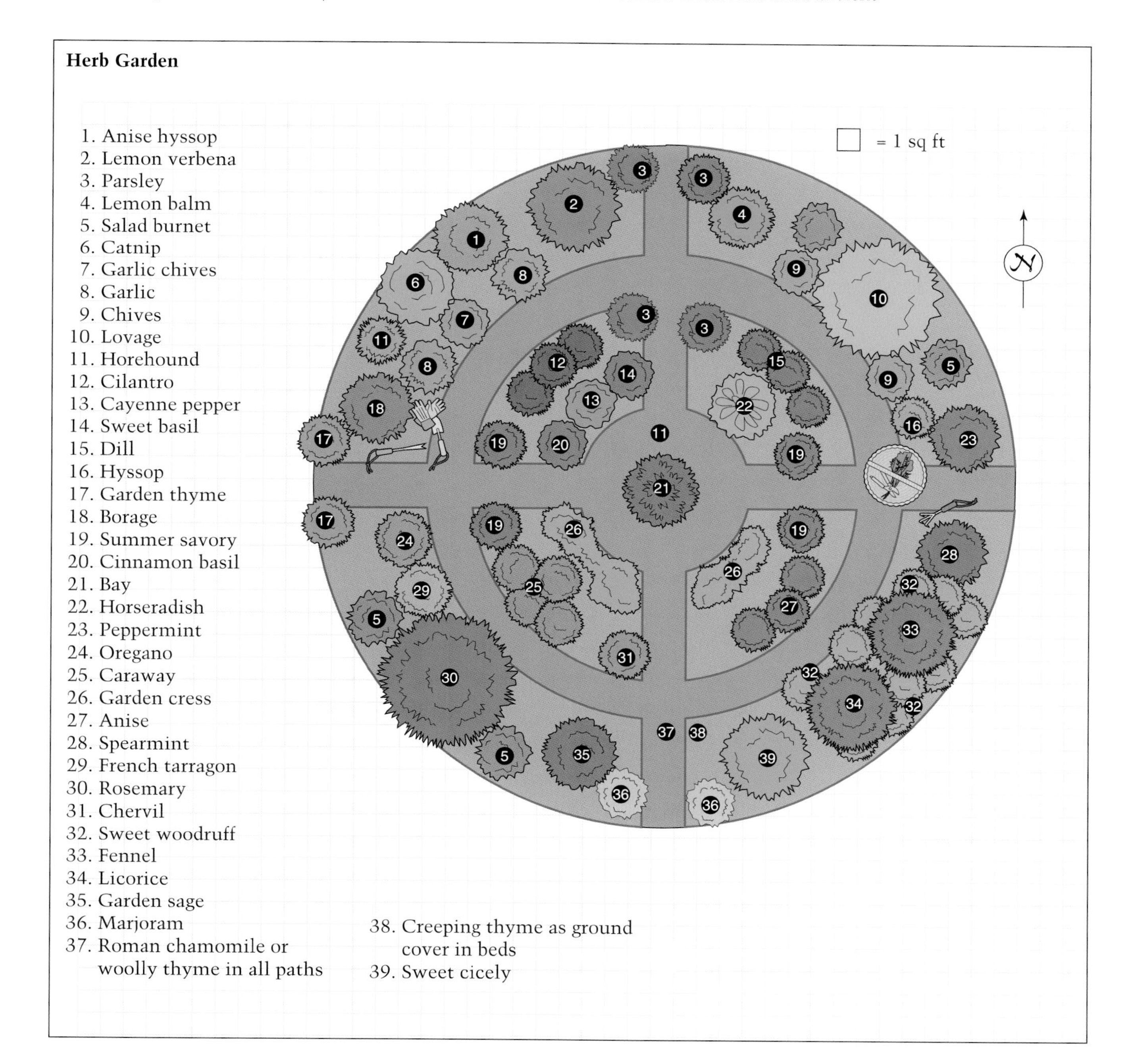

design, one that was used by medieval monks for their apothecary gardens, and by many earlier cultures as well.

The design also embodies another common theme: that of a very rigid formal design that is allowed to soften at the edges as plants overgrow their bounds and spread across its walks. In spite of its formality, the plan is easy to maintain and will require almost no work to keep up, other than harvesting and occasional replanting.

One element that makes the upkeep easy is the use of ground covers in the beds. Ground covers keep weeds down and protect the soil, making cultivating and weeding unnecessary. Their softening influence also adds to the charm of the garden. The ground covers selected are all herbs in their own right. They are soft and low, with the general appearance of a carpet, and fragrant.

The paths are also planted with a ground cover. Roman chamomile has been used for paths, lawns, and even outdoor cushions on raised benches for centuries. It tolerates traffic better than any other plant except turf grass, and releases its chamomile fragrance in response to your step. There is an old folk saying that chamomile doesn't grow well unless it is trod upon.

The center element in this garden is a bay tree. Grow it in a large (16- to 20-inch), ornate pot on a low pad, such as a stepping-stone, in the center of the garden. Bay responds well to shearing, and can be pruned into a topiary—a small tree shape, a globe, or a spiral.

Building the Beds

Lay out these round beds with a large compass. Drive a stake or rod into the center of the garden, tie a loop in one end of a piece of string, and drop the loop over the stake. Tie another loop 8 feet from the stake, and poke a screwdriver through it. Use the screwdriver to scribe a circle with a radius 8 feet in diameter. Scribe other circles at 6½ feet, 5½ feet, and 2 feet from the central stake.

To keep the ground covers in place, edge the beds. Commercially available plastic edging is easiest to install. Dig the paths down a couple of inches, throwing the soil onto the beds. Then install the edging according to the manufacturer's directions.

Planting the Perennials

Most herbs are perennials, and most are probably hardy where you live, but check the plant climate zones to be sure. You can determine your zone from the map on page 91. A few herbs are tender in most zones. The hardy plants can be planted in the ground and left from year to year. Some will live for many years; others have short life spans and will have to be replaced every few years.

It is best to purchase perennials as plants from a nursery. While some can be grown from seed, most start very slowly, and the seed-grown plants may not have the same scent as the purchased ones. If your local nursery does not have the plants you want, order them from one of the companies listed on page 90.

Perennial plants that are tender in your region can be planted in unglazed terra-cotta pots that are *plunged*, or buried in the garden. Cover the drainage hole with copper window screen to keep roots from escaping into the ground while the pot is plunged. Fill the pot with purchased potting soil, plant the herb in it, and set it into the garden with the rim at ground level. Plunged pots can be treated just like garden plants for the summer; the porous pots allow water to pass through, so they don't need special watering.

In the fall, before the first frost, dig up the pot, wash it off, and bring it into a bright, cool room for the winter. An unheated bedroom or attic room is ideal, or it can be placed in a basement if a window is available for light. Water the plant as needed, but let it get quite dry—so the soil an inch below the surface is just slightly moist—before each watering. Don't fertilize over the winter. Move it out to the garden after all danger of frost is past.

You may also want to grow some of the more invasive perennials in pots to limit their spread. Plant the mints and lemon balm in 5-gallon nursery cans (the kind nursery plants come in, either metal or plastic) from which you have cut out the bottom. Allow about 2 inches of the can to protrude above the soil to stop the growth of surface rhizomes (the creeping stems by which these plants spread). A few times each summer, inspect the edges of each can and cut back any rhizomes that are trying to escape its confines. Planting horseradish in a container not only reins in its aggressive root system but makes it easier to harvest the root. Plant it in a 14-inch pot, plunged in the garden.

Horehound, lemon balm, and catnip self-seed easily in some climates and can become pests. This can be avoided by removing the flowers each year before they set seed.

Planting Seeds

Annuals and biennials are usually started from seeds, but you may purchase plants if they are available. Both annuals and biennials die after setting seed. Annuals set seed their first summer; biennials make only foliage during the first summer, then send up a flower stalk and set seed the second summer, usually as soon as the weather warms in spring.

Cilantro, dill, and garden cress are annuals that go to seed quickly, especially when the weather turns warm, so if you want to harvest their leaves, you should plant them several times, in small amounts, as long as the weather is cool. Begin planting in early spring.

Garlic is a perennial that is usually grown as an annual. Plant individual cloves, with the pointed end up, just below the surface of the soil. Space them 3 to 4 inches apart.

Parsley is a biennial that is stimulated to go to seed by cold weather. To have a long period of leafy growth, plant it out in midspring through early summer.

Because the planting times for this garden are so simple, no garden calendar is included. Plant seeds after your frost-free date. Set out perennial plants at any time during the growing season.

Planting the Ground Covers

If you live in zone 7 or warmer, plant chamomile in the walks and creeping thyme in the beds. If you are in a zone colder than zone 7, or in the humid South, use two kinds of thyme for beds and walks.

Both ground covers are most easily established with plugs cut from flats of plants. Cut the flats into 2-inch-square plugs, and plant the plugs 6 inches apart. Planted in the spring, they will fill in by the end of summer. Where it is walked on, the chamomile will stay dense and low. The edging will keep it from growing into the beds.

Harvesting and Maintaining the Herb Garden

Many herbs are grown for their leaves, which are used either fresh or dried. Frequent pinching will keep most plants thick and attractive. If you don't harvest frequently, the plants will be tidier with an occasional shearing. A few kinds, parsley and chives among them, are plants made up of a low rosette of leaves. These should be harvested by cutting single leaves, or a group of leaves, near to the ground. If you want to be able to harvest leaves from a plant over a period of time, never take more than half of it at once.

Harvest the seeds of anise, coriander, dill, and fennel when the leaves of the plants have turned yellow and the fruit gray. Dry the plants on newspapers in a warm, dry spot, then shake the seeds out of their heads. To harvest caraway seed, cut the plants as soon as the seeds begin dropping, and hang them over newspaper (to catch the seeds) in a warm, dry spot. You can leave a plant, or a few seed stems, to sow seeds for the following year's crop.

To harvest plunged horseradish roots, dig up the pot in the fall. Remove the plant from the pot, harvest the root, replant a side shoot for next year, and replace the pot in the garden. Harvest licorice roots after the top dies down in early to late fall, depending on when cold weather arrives. Pieces of root left in the soil will sprout for next year's crop. Dig up garlic in midsummer, after the tops turn brown, and dry the bulbs in the shade for a few days before storing.

Plant List

Herb specialists frequently offer many versions of common herbs. Some of these are closely related species; others are forms that vary in size, texture, flavor, or color from the species. In this list, the most common, or "official," version of the herb is recommended unless there are highly desirable variants—as in the case of basil, thyme, and mint. If you are very fond of a particular herb, you might enjoy trying several varieties.

Because all these herbs are available from most herb specialists, as well as many of the vegetable seed sources, no sources are given. If you can't find the plants you want at your local nursery, request a few herb plant and seed catalogs from the list on page 90.

Anise: Annual
Start from seed in late spring. Only the species is available.

Anise hyssop: Tender perennial, hardy to zone 9
Start from a purchased plant or from seed. Only the species is available.

Basil, cinnamon: Tender perennial, grown as an annual
Start from seed or a purchased seedling in spring to early summer. This variety has no formal name, but is a recognized type of basil with pink flowers and a cinnamon/basil scent.

Basil, sweet: Annual
Start from seed or purchased plant in spring to early summer. Use any large, green-leaved type.

Bay (bay laurel): Tree, hardy to zone 8
Where not hardy, winter indoors. Buy a plant, anytime in mild climates, in spring in colder areas. A compact form with the same flavor is also available.

Borage: Annual
Start from seed, in spring where winters are cold, anytime in mild-winter areas. Only the species is available.

Caraway: Biennial, hardy to zone 4
Start from seed sown in place in early spring. Only the species is available.

Catnip: Perennial, hardy to zone 4
Start from purchased plant in spring or fall. *Nepeta cataria* is the species that drives cats wild. Several other species are available, most of them more attractive and better behaved than *N. cataria,* but without the feline appeal.

Cayenne pepper: Tender perennial grown as an annual
Start in spring from seed or a purchased plant. Although dozens of varieties of hot pepper are available, only one variety is called 'Cayenne'.

Chamomile, Roman: Perennial, hardy to zone 7
Start in spring from purchased flats of plants or seeds. If you want to harvest heads to make tea, be careful not to purchase the nonflowering varieties such as 'Treanague'. Don't use German chamomile *Matricaria recutita*, which isn't suitable as a ground cover.

Chervil: Annual
Start from seed whenever weather is cool but expected to be above freezing. Available as the species or in a curly-leaved form.

Chives: Perennial, hardy to zone 4
Easily grown from seed, or start with plants from a nursery. Plant in clumps. Harvest with scissors, cutting whole leaves from the outside of the clump.

Cilantro (coriander): Annual
Start from seed sown in place whenever the weather is cool and expected to be above freezing for a month or so. Grow the species if you intend to ripen coriander seed. For leaves (cilantro), grow the 'Slow Bolt' or 'Long Standing' varieties.

Cress, garden: Annual
Start from seed whenever the weather is cool but frost is not expected. Several peppery-flavored plants are called cress. The one most commonly grown for seedling use is also known as curled cress or peppergrass. Upland cress and watercress are different plants.

Dill: Annual
Start from seed in mid to late spring. Pick the leaves as needed for the kitchen. The species is usually grown, but several varieties are available, including some slow-to-bolt varieties, useful if you are more interested in dillweed than in seeds.

Fennel: Perennial, hardy to zone 6
Start from seed sown in place in early spring. Sweet fennel is available in the common green form and a more ornamental bronze variety. One variety, Florence fennel, has thickened leaf bases and is used as a vegetable, not an herb.

Garlic: Perennial grown as an annual
Start from purchased sets in fall or late winter to early spring. There are many varieties; look for one that is said to be productive where you live, or one that is widely adapted, such as 'Italian Purple Skin' (Shepherd's). Elephant garlic is a different plant (see page 76 for more information about it).

Garlic chives: Perennial, hardy to zone 5
Start from seed in spring or purchased plants in spring or fall. This is not a variety of chives, but a different member of the onion tribe.

Horehound: Perennial, hardy to zone 4
Start from a purchased plant in spring or fall. The species is most commonly grown.

Horseradish: Perennial, hardy to zone 2
Start from a root cutting in the spring. The species is most commonly grown.

Hyssop: Perennial, hardy to zone 3
Start from seed started indoors in early spring or from a purchased plant in spring or fall. The species is most commonly grown.

Lemon balm: Perennial, hardy to zone 4
Start from a purchased plant in spring or fall or from seed. Several variegated varieties are available with the same flavor as the species, and there is also a lime-scented variety.

Lemon verbena: Perennial, hardy to zone 8
Start from a purchased plant in spring. The species is most commonly grown. In cold-winter areas, it may be hard to find locally.

Where it is not hardy, grow in a 12- to 14-inch pot. Pinch or shear the plant a couple of times during the summer. It will probably drop its leaves when brought inside for the winter, but will grow a new set in the spring when you return it to the garden.

Licorice: Perennial, hardy to zone 9
Cover with straw mulch in colder winters. Start from root cuttings in fall or from a purchased plant in spring. The species is the source of commercial licorice flavor.

Lovage: Perennial, hardy to zone 3
Start from purchased plants in spring or fall. The species is most commonly grown.

Marjoram (*Origanum marjoranum*): Perennial, hardy to zone 9
Start from a purchased plant in spring. Sweet marjoram is usually grown as an annual in zone 8 and colder areas. The species is most commonly grown.

Mint, Peppermint: Perennial, hardy to zone 5
Start from purchased plants in spring or fall. Dozens of other mints are available, each with its own flavor.

Mint, Spearmint: Perennial, hardy to zone 5
Start from purchased plants in spring or fall. This is the most commonly grown mint.

Oregano: Perennial, hardy to zone 5
Start from a purchased plant in spring or fall. Many species and varieties are sold as oregano. Although the name on the plant may not be reliable, some that usually have a good flavor are *Origanum vulgare hirtum, O. onites*, and *O. heracleoticum*. Your best bet is to purchase plants, letting your nose guide you to one with a scent you like.

Parsley: Biennial usually grown as an annual
Best started from purchased plants, set out in spring to early summer. Many varieties are available. The flat-leaf or plain-leaf type has a richer parsley flavor. The curled-leaf varieties are more ornamental as garnishes.

Rosemary: Shrub, hardy to zone 8
Several prostrate forms are available, as well as a hardy variety ('Arp') that has been grown outside to zone 6. In zones where rosemary is not hardy, you will need to plunge it (see page 52). For plunging, select the species and shear it once a year to keep it compact. Plunged plants will not get as large as plants grown in the ground in mild climates. Where rosemary must be plunged, it will not grow as large as shown on the plan; add a couple more types of sage and oregano to take up the extra space.

Sage, garden: Perennial, hardy to zone 5
Start from a purchased plant in spring or fall. Buy the species or varieties such as 'Holt's Mammoth' (large leaves), 'Aurea' (golden leaves), or 'Tricolor' (green, pink, and white striped leaves). All have the same sage flavor. There are many other sage species, some used in cooking, others as landscape plants.

Salad burnet: Perennial, hardy to zone 3
Start from seed in spring or fall. Only the species is available.

Savory (summer savory): Annual
Start from seed in spring. Perennial, or winter, savory is similar, but it has a stronger, sharper flavor.

Sweet cicely: Perennial, hardy to zone 3
Start from purchased plants in fall or spring. Only the species is available.

Sweet woodruff: Perennial, hardy to zone 3
Start from purchased plants in spring or late summer. This is the only plant in the plan that grows better in shade than in sun.

Tarragon, French: Perennial, hardy to zone 4
Short-lived where winters are mild. Plants are completely dormant in winter, even in mild-winter areas. Start from purchased plants in spring. True French tarragon is propagated only by plants grown from cuttings; seed will produce plants that do not have the characteristic flavor of French tarragon.

Thyme, creeping (*Thymus praecox arcticus*): Perennial ground cover, hardy to zone 5
Start from purchased plants in spring. Dozens of species and varieties of ground cover thymes are available. *Thymus praecox* 'Albus' is a white-flowered variety of wild thyme that will make a good ground cover in the herb beds.

Thyme, garden (*Thymus vulgaris*): Perennial, hardy to zone 5
Start from purchased plants in spring or fall. This species is the one most often used in cooking. As with oregano, there are many varieties, including English thyme, French thyme, variegated thyme, and lime thyme; let your nose guide you to the plant you want to grow.

Thyme, woolly (*Thymus pseudolanuginosus*): Perennial ground cover, hardy to zone 5
Start from purchased plants in spring. Use for a ground cover if you are too far north to grow chamomile in the paths. Woolly thyme takes more foot traffic than most creeping thymes, and looks different enough from creeping thyme to make a nice contrast. It is not usually used as a culinary herb.

A PATIO GARDEN

If sitting on a patio among your vegetables, herbs, and flowers appeals to you, you will like this plan, which makes your garden just the place for a summer breakfast, a chat with friends, or a quiet moment of relaxation. The patio area is big enough for a small table and chairs. Practical as well, this plan also offers you a good choice of garden delicacies.

An ample strawberry bed gives you plenty of plants for nibbling or for shortcake. And two rhubarb plants will be enough for many a rhubarb (or rhubarb-strawberry) dessert. There are all of your summertime favorites: tomato, pepper, eggplant, snap beans, summer squash, and cucumber. And there's a full assortment of salad fixings.

Some of the tomato varieties in this garden represent a newer type of tomato plant, called a dwarf indeterminate. They are short, yet they continue bearing fruit as long as many taller varieties do. These plants need no staking or

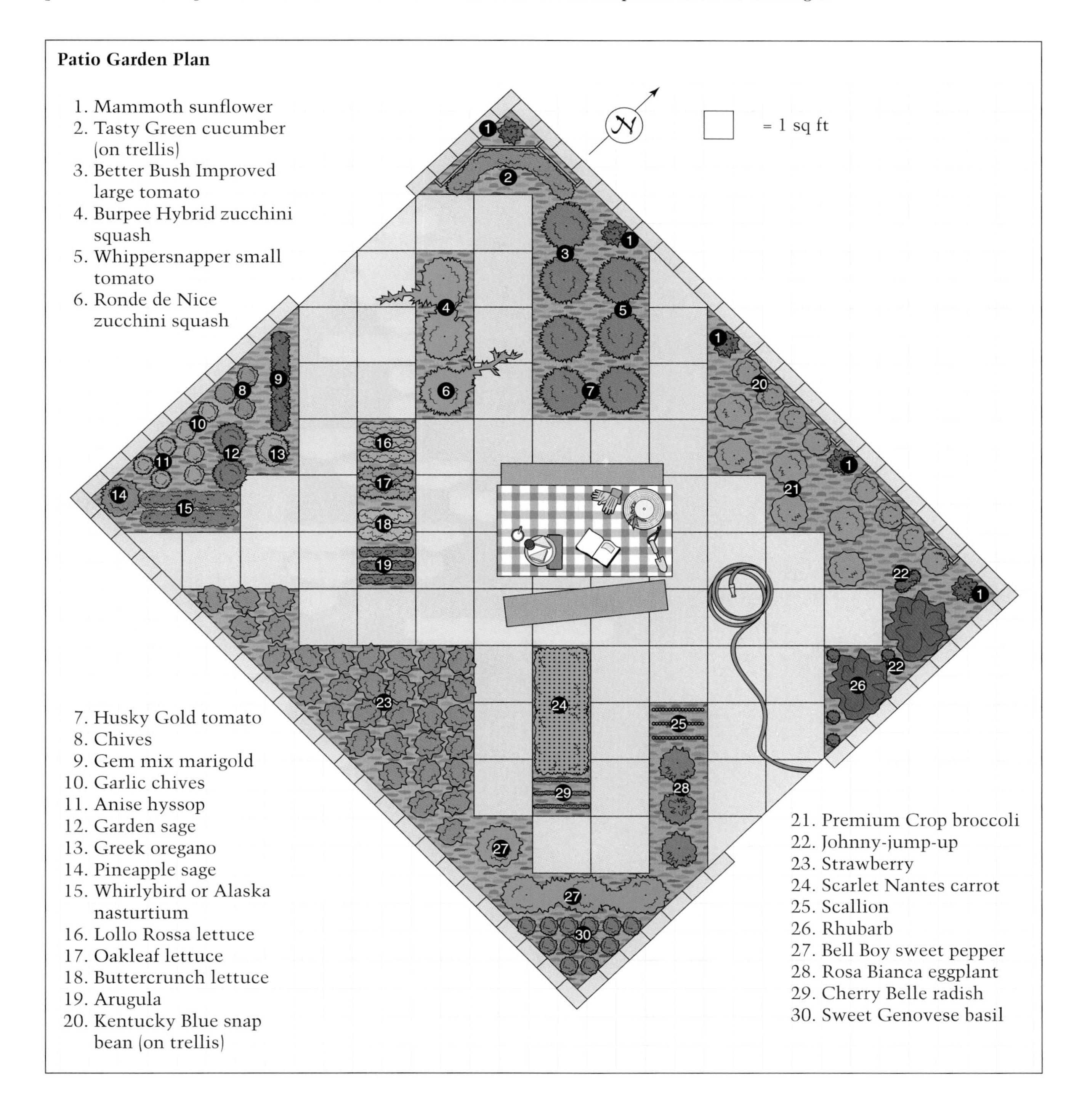

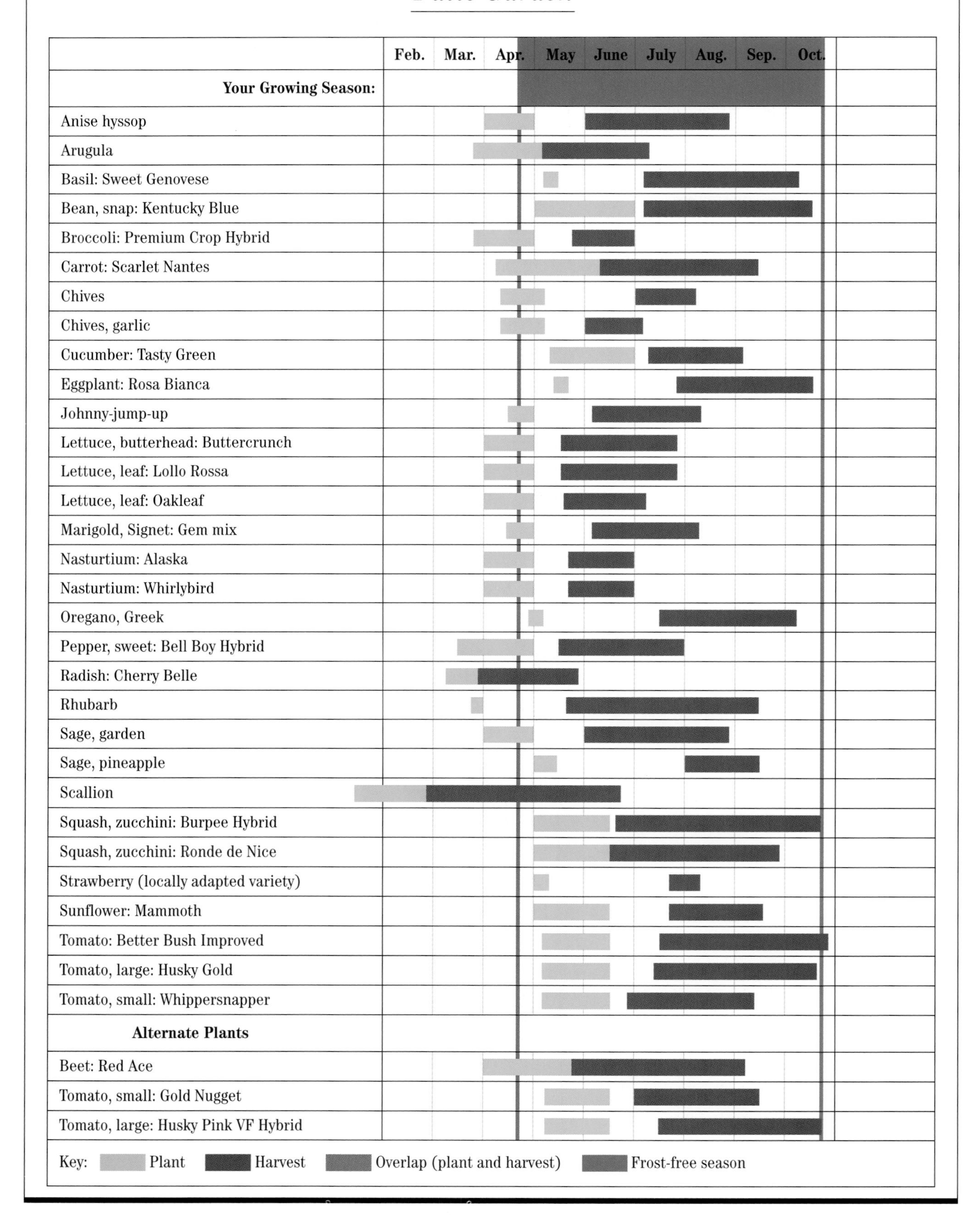
Patio Garden
Feb.
Mar.
Apr.
May
June
July
Aug.
Sep.
Oct.
Your Growing Season:
Anise hyssop
Arugula
Basil: Sweet Genovese
Bean, snap: Kentucky Blue
Broccoli: Premium Crop Hybrid
Carrot: Scarlet Nantes
Chives
Chives, garlic
Cucumber: Tasty Green
Eggplant: Rosa Bianca
Johnny-jump-up
Lettuce, butterhead: Buttercrunch
Lettuce, leaf: Lollo Rossa
Lettuce, leaf: Oakleaf
Marigold, Signet: Gem mix
Nasturtium: Alaska
Nasturtium: Whirlybird
Oregano, Greek
Pepper, sweet: Bell Boy Hybrid
Radish: Cherry Belle
Rhubarb
Sage, garden
Sage, pineapple
Scallion
Squash, zucchini: Burpee Hybrid
Squash, zucchini: Ronde de Nice
Strawberry (locally adapted variety)
Sunflower: Mammoth
Tomato: Better Bush Improved
Tomato, large: Husky Gold
Tomato, small: Whippersnapper
Alternate Plants
Beet: Red Ace
Tomato, small: Gold Nugget
Tomato, large: Husky Pink VF Hybrid
Key:
Plant
Harvest
Overlap (plant and harvest)
Frost-free season

caging and are tidier and more attractive than the taller types. They are available in red and yellow (shown in the plan), and in a pink-fruited variety that is listed as an alternate.

A bed in one corner of the patio provides herbs for cooking and also some dual-use flowers—in addition to being good cut flowers, all are edible. Try nasturtium or sage flowers in a tossed salad, anise hyssop flowers in a fruit salad, or bright marigold petals sprinkled on a bowl of cream soup (see page 27 for more on edible flowers). Elsewhere in the garden, you can slip in a few Johnny-jump-ups, Gem marigolds, or nasturtiums to brighten the corners of beds.

Finally, in the rear, some tall sunflowers will lean over your garden, turning their golden heads to the summer sun. When they are ripe, harvest the heads for seeds. Roast these for winter snacks, or put the heads out in winter and watch the birds eat their fill.

About the Plan

Setting pavers on the diagonal makes this garden less formal and gives it more dynamic interest than it would have if the pavers and beds all ran parallel to the edges of the patio. The beds surround the seating area prettily with about 200 square feet of garden. The beds are narrow enough, and have access from enough directions, that harvesting is easy. A 4-foot-wide path allows wheelbarrows or garden furniture to be moved in and out of the central area.

Use 2-foot-square pavers to create this garden. They could be flagstone, plain concrete, or exposed-aggregate concrete. Set the pavers in a 2-inch layer of sand, with their edges touching. You will have to cut a few half-pavers for this plan.

When you build this garden, think about how much access you will need from the outside on each side, and about the boundary between each side and the neighboring lawn or other features. Be careful not to block access to the outside of the strawberry beds or the ends of the two wide paths that lead into the patio. You could lay a border of narrow pavers, bricks, or wooden landscape timbers, set flush with the ground. Such a border would contain your beds on the outside and, if a lawn is the adjoining surface, would serve as a mowing strip, making the edge of your lawn much easier to maintain.

This small garden won't provide you with as many vegetables as some of the other plans in this book, but it will keep some fresh salads and vegetables on your plate for most of the season. It is a dooryard garden, a small garden close to the kitchen door that the cook can visit for some fresh herbs, salad makings, or a mess of beans.

Plant List

Anise hyssop (licorice mint)
With its sweet, anise-scented leaves and spikes of small purple flowers, this plant is very pretty in the garden. Use leaves for tea or in fruit salad, flowers to decorate a fruit salad or in a bouquet. Ready in 90 days. Abundant Life, Cook's, Pinetree, Southern Exposure, Territorial.

Arugula
Also known as rocket, roquette, and rucola. Ready in 45 days. Widely available.

Basil: Sweet Genovese (Genovese, Genova Profumatissima)
Or get any basil designated "large" or "large-leafed Italian." Ready in 60 days from seed. Cook's Garden, Johnny's, Nichols, Seeds of Change, Shepherd's.

Bean, pole, snap: Kentucky Blue
These are sweeter than 'Kentucky Wonder'. The plants bear an abundant crop of uniform 7-inch pods. Ready in 65 days. Pole beans bear over a longer period than bush beans. Widely available.

Broccoli: Premium Crop Hybrid
Makes good heads later into summer heat than most broccolis. Ready in 62 days. Widely available.

Carrot: Scarlet Nantes
Cylindrical, about 6 inches long. Ready in 70 days. Widely available.

Chives
The narrow, spiky leaves of chives have an onion flavor, and are usually added at the end of cooking. The lavender flowers are also edible. This perennial herb is best grown using purchased plants.

Cucumber: Tasty Green 26 Hybrid
This is one of the "burpless" types with mild flavor and a thin skin. It resists powdery and downy mildew. Ready in 62 days. Burpee, Field's, Gurney's, Nichols, Park.

Eggplant: Rosa Bianca
An Italian heirloom variety with small, pink fruit. Ready in 75 days. Shepherd's.

Garlic chives
Start from seed in spring or from purchased plants in spring or fall. They offer a unique flavor and lovely stems of edible white flowers. 150 days to first cutting. Widely available.

Johnny-jump-up
Small tricolor (purple, lavender, and yellow) blooms are easy to grow from seed. Eat the whole flowers. Widely available.

Lettuce, butterhead: Buttercrunch
Forms a loose head of smooth, dark green leaves. Slow to bolt. Ready in 55 to 65 days. Widely available.

Lettuce, leaf: Lollo Rossa
Compact, heat-tolerant plants with frilly leaves that are pale green in the center, tinged ruby red at the tips. Ready in 50 days. Cook's Garden, Shepherd's, Vermont Bean Seed.

Lettuce, leaf: Oakleaf
Frilly, lobed, light green leaves have good flavor and stand up well to heat. Ready in 40 to 50 days. Widely available.

Marigold, Signet: Gem mix
All marigolds (*Tagetes* species) are edible, but Signet marigolds (*T. tenuifolia*) have the best flavor. The Gem mix contains seeds of orange, pale yellow, and gold flowers, or you can buy single colors. Pull out petals and add to salads. Cook's Garden, Shepherd's.

Nasturtium: Alaska (Tip Top Alaska)
Choice nasturtium selection with green-and-white variegated leaves and flowers in shades of yellow and orange. The edible leaves and flowers have a spicy flavor. Plants are compact. Ready in 60 days. Burpee, Cook's, Nichols, Pinetree, Shepherd's.

Nasturtium: Whirlybird
A compact nasturtium with flowers in 6 bright colors. Flowers and leaves have a spicy flavor. Ready in 60 days. Park. Cook's Garden carries a similar 'Tip Top Mix'.

Onion, scallion
Purchase sets in the spring and plant them out to grow scallions. Ready in 30 days. Widely available.

Oregano, Greek
Grow this popular perennial herb to flavor salad dressings. Buy a plant locally, but rub the surface of a leaf and smell it to see if it has a good scent before you buy it.

Pepper, sweet: Bell Boy Hybrid
A large green bell pepper that ripens to red. Resists tobacco mosaic virus. Ready in 62 to 70 days. Widely available.

Radish: Cherry Belle
Round and bright red. Ready in 22 days. Widely available.

Rhubarb: Locally adapted variety
Not well adapted to areas with very mild winters, where it may not survive past one year or may have stalks that are less red. Buy roots locally. Where rhubarb is not adapted, consider artichokes instead.

Sage, garden
Sage comes in several color and size variations, as well as the species. All are perennial and have the same sage flavor. Buy plants locally. Widely available.

Sage, pineapple
This plant has a light pineapple scent that will nicely season iced tea or fruit salad. The vivid red flowers, borne in late summer, are edible and attract hummingbirds. Buy a plant locally.

Squash, summer, zucchini: Burpee Hybrid
Vigorous, compact plants bear many medium-green zucchinis with thin skins and good flavor. Ready in 50 days. Burpee.

Squash, summer, zucchini: Ronde de Nice (Round French)
A pale green, round zucchini that is a European heirloom. Harvest squash at 1 to 5 inches in diameter. Ready in 45 days. Cook's Garden, Seeds of Change, Shepherd's, Vermont Bean Seed.

Strawberry: Locally adapted variety
Strawberry varieties are developed for specific regions, and plants are best purchased locally.

Sunflower: Mammoth (Mammoth Russian)
Grow this old favorite, or try one of the newer hybrids. Ready in 110 days. Widely available.

Tomato, large: Better Bush Improved VFF Hybrid
One of the newer long-bearing types that need no staking and have a tidy appearance. The red fruit is 3 to 4 inches across, meaty, and flavorful. Ready in 72 days. Park.

Tomato, large: Husky Gold VF Hybrid
Golden tomatoes and dark green leaves on long-bearing plants. Its short stature makes this variety a good choice for a patio or containers. The fruits average 8 ounces each and have a mild, sweet flavor. Ready in 70 days. Nichols, Tomato Growers Supply.

Tomato, small: Whippersnapper
Extra-early cherry tomatoes ripen on these determinate plants. The flavorful fruit is pinkish red, oval. Also suited to hanging baskets. Ready in 52 days. Johnny's.

Alternate Plants

Beet: Red Ace
These beets develop fast and stay tender and sweet for a long time. They have good disease tolerance. Ready in 53 days. Field's, Park, Pinetree, Territorial, Vermont Bean Seed.

Tomato, small: Gold Nugget
These compact determinate plants bear 1-inch-diameter gold cherry tomatoes with a rich, sweet flavor that ripen without cracking or falling off of the vine. Most fruits are seedless. Ready in 60 days. Johnny's, Nichols, Pinetree, Territorial, Tomato Growers Supply.

Tomato, large: Husky Pink VF Hybrid
A pink-fruited version of the long-bearing but short tomato plant. Dark green, quilted leaves set off mild-flavored, 5-ounce pink fruits. Ready in 70 days. Nichols, Tomato Growers Supply.

CONTAINER VEGETABLE GARDENS

If the only place you have for a garden is on a concrete surface, or on a deck, porch, roof, or even a balcony, you can still have fun growing vegetables and fruits. As with a garden in the ground, the preparation is important, but once you have a container garden in place, it can be easy and productive. The setup can be inexpensive and simple, or as elaborate as you wish.

For containers, you can use recycled buckets, boxes, or cans; purchase new pots and boxes at a nursery; or construct your own custom containers to fit the available area. Just be sure that the containers have small drainage holes in their bottoms.

Minimum Container Depth for Some Vegetables, Herbs, and Edible Flowers

Plant	Depth
Arugula	6–8"
Basil	8"
Bush beans	8–10"
Pole beans	10–12"
Small beets*	8"
Round carrots	6"
Carrots to 8 inches*	12"
Chard	10–12"
Cottage pinks	8"
Cucumbers	12"
Eggplant	10–12"
Garlic chives	10"
Johnny-jump-up	8"
Lettuce	8"
Miniature leaf lettuce	6"
Signet (Gem) marigolds	8"
Mizuna	8"
Nasturtiums	8"
Scallions	8"
Oregano	10"
Parsley	10"
Bush and pole peas	12"
Standard peppers	12"
Compact peppers	8"
Standard radishes	6"
Long radishes	8"
Summer squash	24"
Strawberries, Alpine and standard	8"
Tomatoes, indeterminate (tall)	15–24"
Tomatoes, determinate (short)	1–18"

*Don't attempt to grow large beets, or carrots longer than 8 inches, in containers.

You can produce a surprising amount of food in a small space. Some of the best crops for container gardens are lettuce, radishes, carrots, tomatoes, peppers, edible flowers, and herbs, but many others are good as well.

In addition to being fun and productive, a container garden can be beautiful. An arrangement of edible crops in containers is often charming just for what it is. Judiciously placed containers of flowers will add considerably to that charm. And a container vegetable garden can be designed to work as an ornamental garden, using a unified design, plus trellises, benches, and other garden features.

Container Size and Weight

To be sure of success, make sure your containers are the right size for the crops you hope to grow. Only with an adequate volume of soil can your plants reach their full size. The box at left gives minimum depths.

You will probably want to include containers of different depths. Lettuce is fine in an 8-inch-deep container, while standard tall tomato plants need at least a foot, and will be taller if they have as much as 24 inches in which to spread out roots and get comfortable.

Wooden containers should be raised on small wooden or ceramic blocks to slow the decay of the container bottom. And all containers should be placed on small blocks if they are standing on a wooden deck, or the portion of the deck underneath the container will decay.

Plants in containers must be grown in a special potting mix. If you try to grow container vegetables in regular garden soil, you are likely to find that they always look underwatered or, if the soil is mostly clay, that it bakes hard when you miss a watering and refuses to let the water back in. Also, garden soil is quite heavy. If your garden is on a rooftop or raised porch, you will want to keep the mix as light as possible. Select a potting mix by weight, looking for one that contains no sand. A mix based on peat moss and perlite is the lightest.

If your garden is on a rooftop or balcony, consider the overall weight of the garden. While most balconies are designed to carry quite a bit of weight, a very large garden might be too much. Most rooftops are not designed for the weight of a container garden. If you have doubts, check with a structural engineer.

About the Plans

The first two container gardens are variations on a single theme of high food production in a small space. One is designed for a deck adjacent to a back door, with a wide path to stairs at the edge of the deck. The other is intended for a balcony or enclosed deck, with no access except the house door.

The choice of crops in all of the plans is similar, consisting mostly of those suited for making salad. It is assumed that the gardener will replant areas when quick crops such as lettuce or radishes have been harvested. If the weather turns too hot for these crops, set out plants of malabar or some summer annual flowers, then plant heat-sensitive crops again when cooler fall weather approaches.

These plans call for large boxes made of wood to be used as planting containers. Such containers are easier to maintain than small ones because the larger volume of soil stays moist longer, meaning less frequent watering.

In two of the gardens there is a box containing full-sized cucumbers. Train them up a trellis attached to the back of the box.

The tomatoes in two of the container gardens are dwarf indeterminate types. While they are short, these varieties bear heavily over a long period, as do regular indeterminate types. They are more upright and wind resistant than other tomatoes, but they should still be trellised in windy sites. Use plastic garden tape to tie them up as they grow.

In general, whenever you can trellis a container plant, do so. This often will allow you to take advantage of the production level of full-sized vegetable varieties.

A miniature fruit tree has been included in the balcony garden. Miniatures are genetic dwarfs, smaller even than the usual dwarf fruit trees. You can grow one in a half barrel. If it is not hardy in your area, put it on rollers and roll it indoors during the coldest part of winter.

The third garden, featuring a bench under an arbor, has been planned for appearance first and production second. It can be made on a balcony or small deck. Choose the purple-tinged 'Trionfo Violetto' or the red-flowered 'Scarlet Runner' pole beans to cover the arbor with a dramatic display. Next to the bench are

A Trellis Box

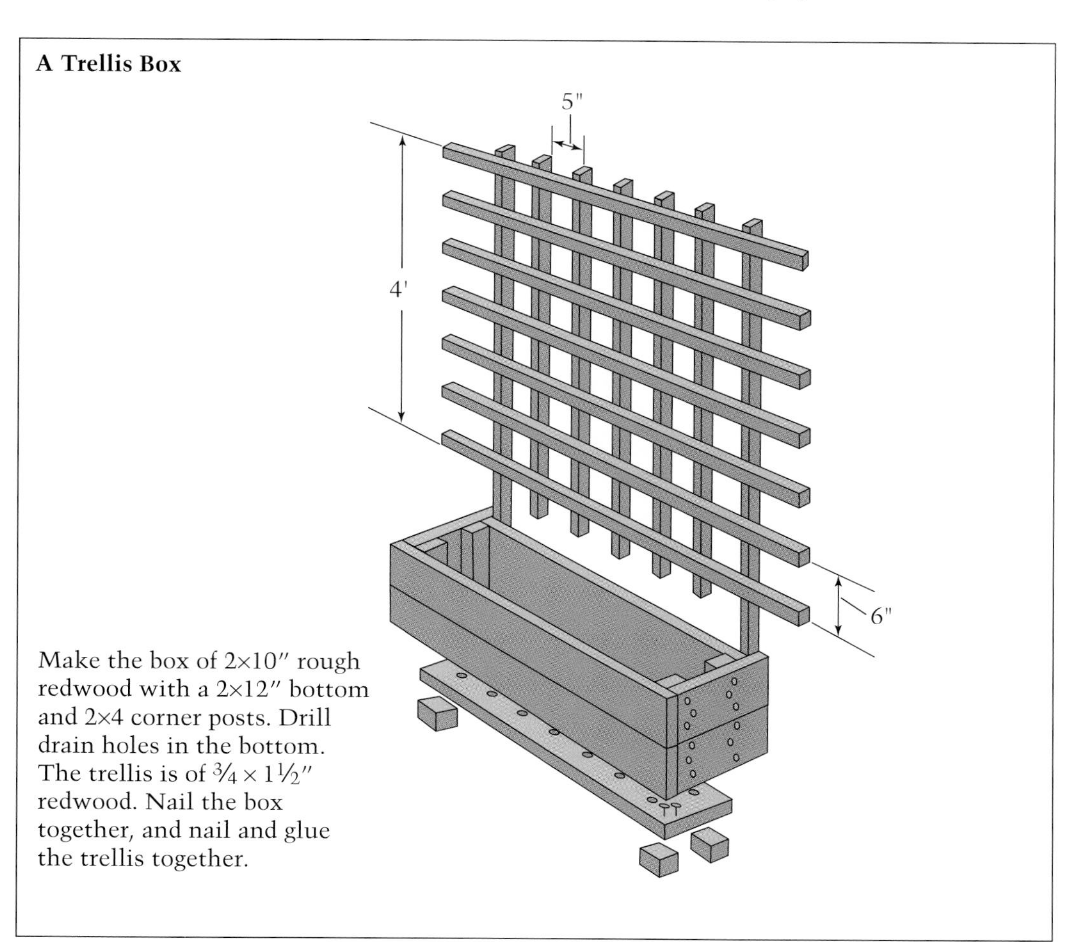

Make the box of 2×10″ rough redwood with a 2×12″ bottom and 2×4 corner posts. Drill drain holes in the bottom. The trellis is of $3/4 \times 1\frac{1}{2}$″ redwood. Nail the box together, and nail and glue the trellis together.

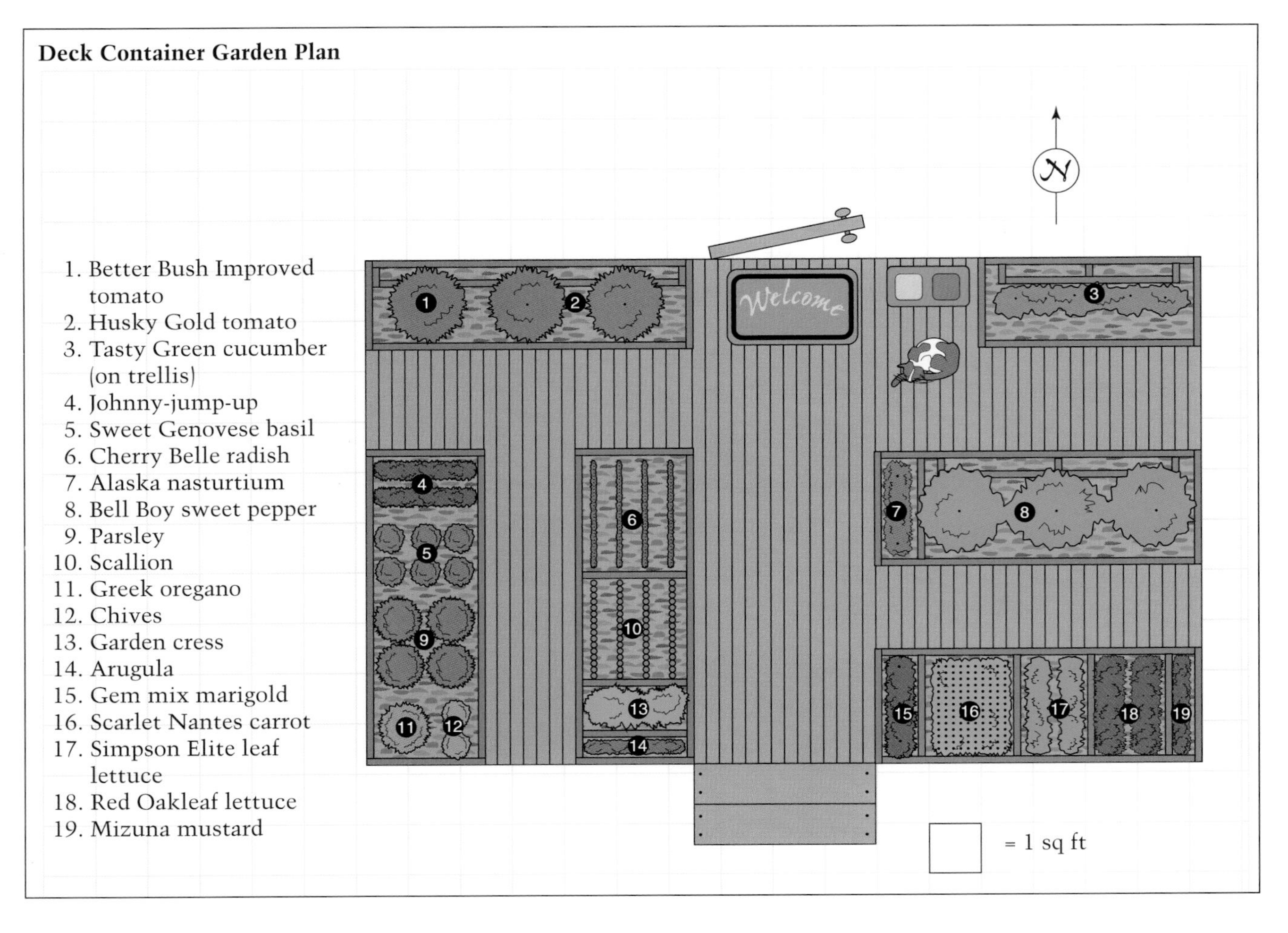

pots of Alpine strawberries for nibbling, and parsley for its clean, green foliage. Both can take the partial shade cast by the beans.

This plan makes ample use of dwarf fruiting crops that do well in small pots. The four pots hanging from the trellis contain miniature "patio" type tomatoes. Two are 'Toy Boy', which makes plenty of 1½-inch red fruit on plants that hang down 14 inches. The other two are 'Floragold Basket', for yellow cherry-sized fruit on similar plants. Both can be planted more than one to a pot, for a fuller appearance.

None of these gardens provides a large amount of food, but you will be surprised at how much you can grow in a small area without benefit of soil!

Plant List

Arugula
Also known as rocket, roquette, and rucola. Ready in 45 days. Widely available.

Basil: Sweet Genovese (Genovese, Genova Profumatissima)
Or get any basil designated "large" or "large-leafed Italian." Ready in 60 days from seed. Cook's Garden, Johnny's, Nichols, Seeds of Change, Shepherd's.

Bean, pole, snap: Trionfo Violetto
Dark green leaves with purple veins, deep lavender flowers, and deep purple snap beans make this one of the prettiest plants you can grow in your vegetable garden, and the beans are great eating as well. Pods cook to green. Ready in 62 days. Cook's Garden, Johnny's.

Carrot: Scarlet Nantes
Cylindrical, about 6 inches long. Ready in 70 days. Widely available.

Carrot: Little Finger
This miniature or "baby" carrot variety is 3 to 5 inches long and ½ inch in diameter. Ready in 65 days. Field's, Gurney's, Pinetree, Vermont Bean Seed.

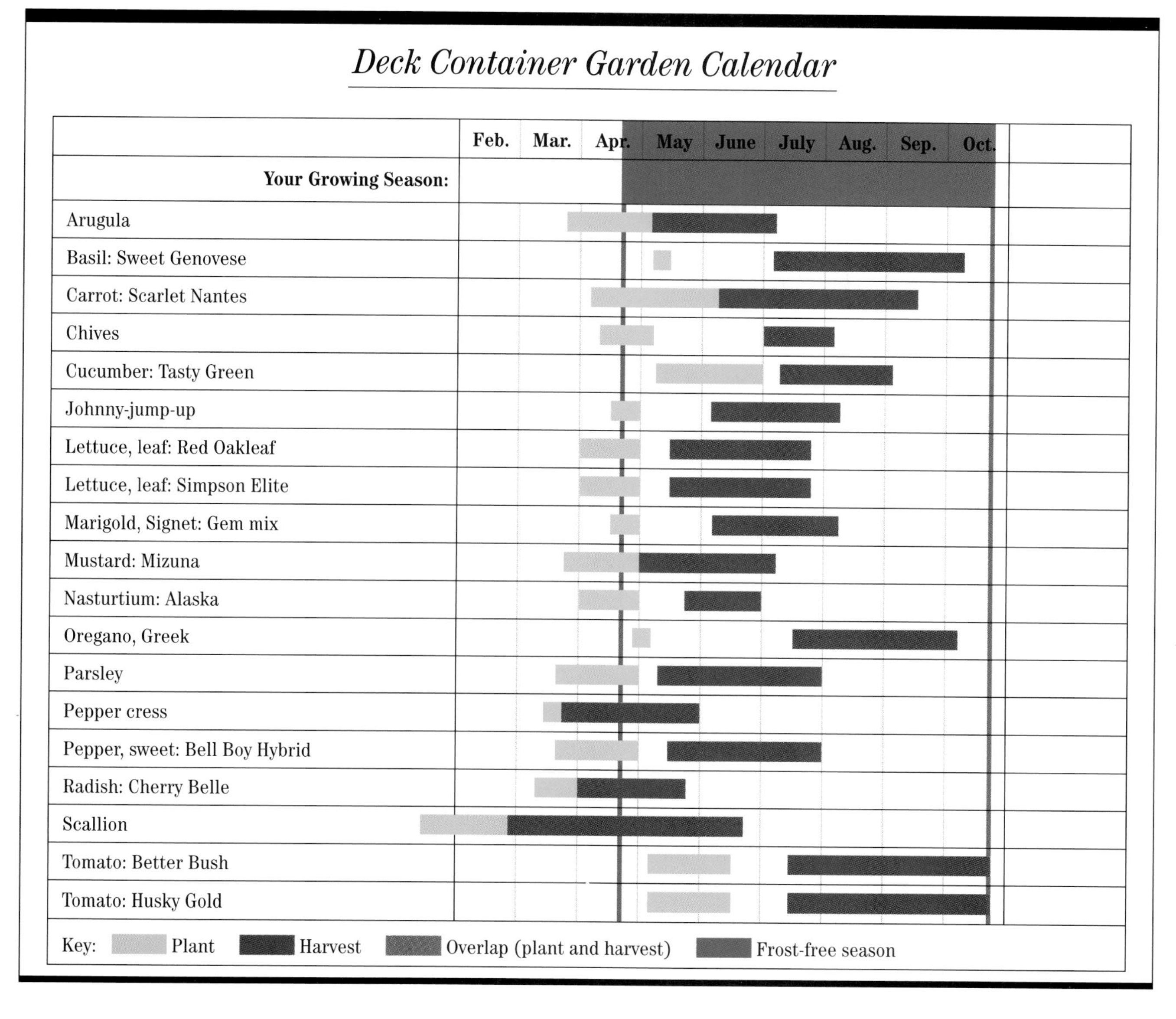

Chives
A perennial herb grown for its narrow, tubular, onion-flavored leaves and edible lavender flowers. Purchase plants locally.

Cottage pink
This lovely perennial makes 6- to 8-inch-high mats of edible flowers in white, pink, red, and bicolors. It can be grown from seed, although it is more often grown from divisions or from nursery starts.

Cress, garden (peppergrass)
A fast-growing annual that is at its best in cool weather, cress adds a peppery tang to salads, omelets, or sandwiches. Try broadleaf or curly cress. Ready in 10 to 30 days. Cook's Garden, Johnny's, Shepherd's.

Cucumber: Salad Bush Hybrid
The maximum vine length is about 22 inches, but it produces plenty of 8-inch slicers. The plants tolerate downy and powdery mildew, target leaf spot, and cucumber mosaic virus. Ready in 57 days. Burpee, Field's, Pinetree, Shepherd's.

Cucumber: Tasty Green 26 Hybrid
This is a delicious, burpless type that is also featured in several of the other gardens in this book. Any standard, long-vined cucumber may be grown in a container if the container is deep enough. It is highly advisable to trellis the plants so that they take up less ground space. Ready in 62 days. Burpee, Field's, Gurney's, Nichols, Park.

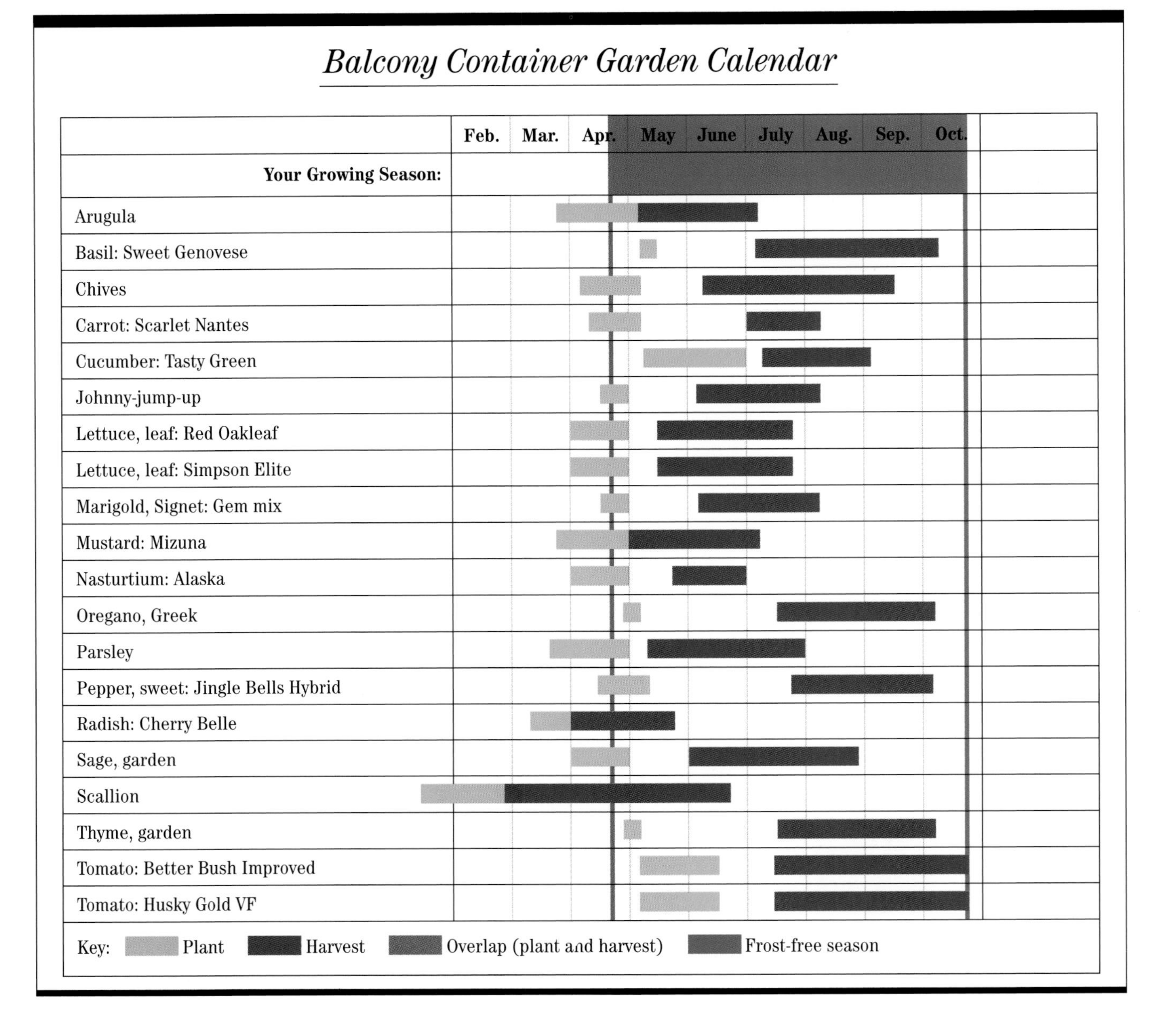

Eggplant: Asian Bride
Very productive and pretty, these 2-foot-tall plants have elongated fruits that are white streaked with lavender. Ready in 70 days. Shepherd's.

Eggplant: Easter Egg Hybrid
Looking like oversized white eggs, these have good flavor. Each 23-inch-tall plant will bear about a dozen fruits. They should be eaten before they ripen to yellow. Ready in 52 days. Pinetree.

Garlic, for leaves
You can plant some garlic cloves, let them grow for a couple of months, and harvest whole young plants, or just the outer leaves, to use fresh in cooking. These can be replanted frequently for an ongoing supply. Widely available.

Garlic chives
A perennial herb with the flavor of both garlic and chive. The flat leaves are good in salad or cooking, and the pretty white flowers are also edible. Harvest by the leaf, or use whole plants. Widely available.

Johnny-jump-up
Easy to grow from seed. Eat whole flowers. Widely available.

Balcony Container Garden Plan

1. Better Bush Improved tomato
2. Husky Gold tomato
3. Jingle Bells sweet pepper
4. Tasty Green cucumber (on trellis)
5. Miniature fruit tree
6. Alaska nasturtium
7. Red Oakleaf lettuce
8. Simpson Elite lettuce
9. Mizuna mustard
10. Johnny-jump-up
11. Sweet Genovese basil
12. Parsley
13. Cherry Belle radish
14. Arugula
15. Gem mix marigold
16. Scallion
17. Scarlet Nantes carrot
18. Chives
19. Garden thyme
20. Greek oregano
21. Garden sage

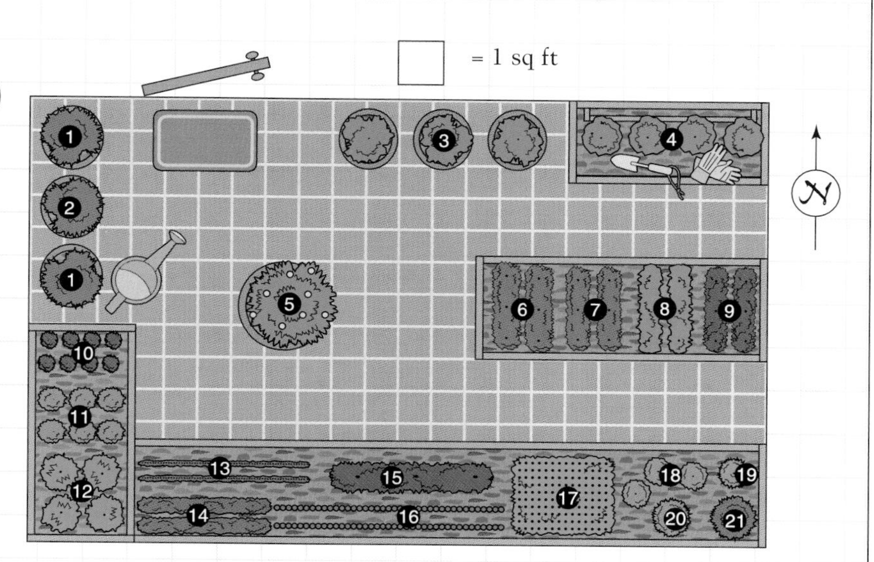

Ornamental/Edible Container Garden Plan

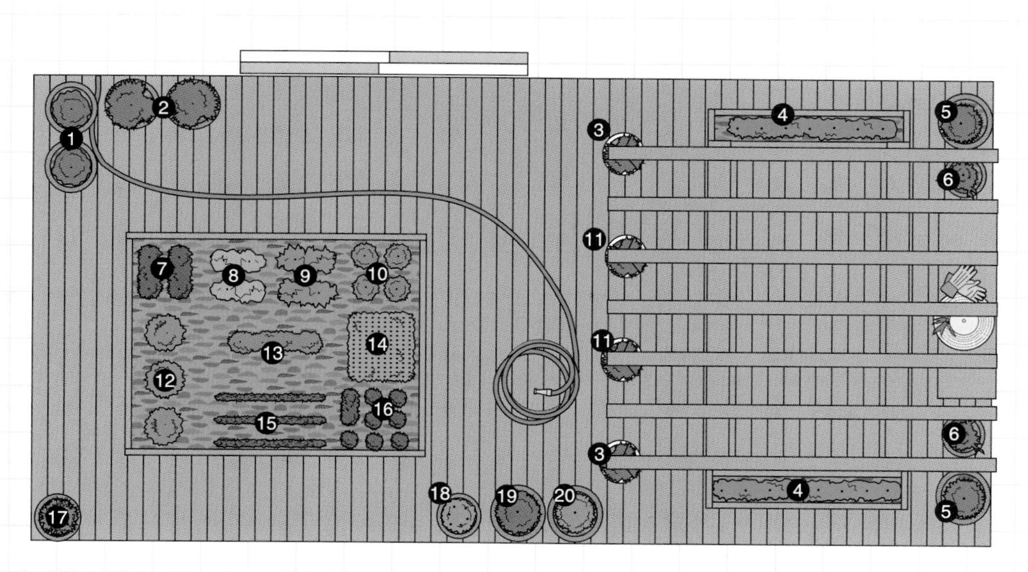

1. Asian Bride eggplant
2. Easter Egg eggplant
3. Toy Boy tomato in basket
4. Trionfo Violetto snap bean
5. Parsley
6. Ruegen Improved Alpine strawberry
7. Gem mix marigold
8. Buttercrunch butterhead lettuce
9. Red Sails leaf lettuce
10. Cottage pink
11. Floragold Basket tomato in basket
12. Sweet Pickle pepper
13. Salad Bush cucumber
14. Little Finger carrot
15. French Breakfast radish
16. Johnny-jump-up
17. Garlic
18. Garlic chives
19. Garden sage
20. Greek oregano

= 1 sq ft

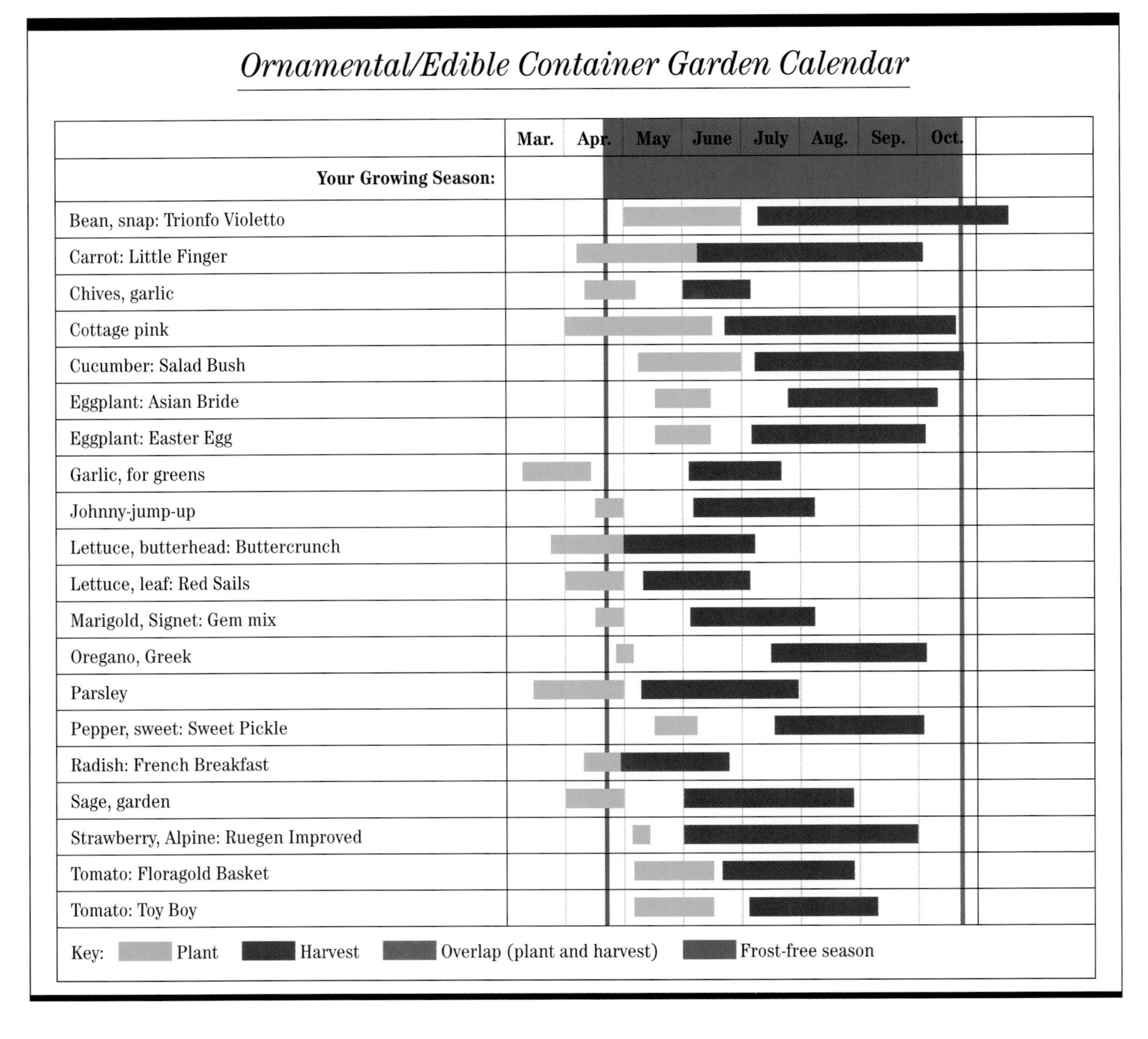

Lettuce, butterhead: Buttercrunch
Forms a loose head of smooth, dark green leaves. Slow to bolt. Ready in 55 to 65 days. Widely available.

Lettuce, leaf: Red Oakleaf (Red Salad Bowl)
Elegant deep maroon leaves. Ready in 50 days. Cook's Garden, Johnny's, Shepherd's.

Lettuce, leaf: Red Sails
Crinkled, bronze-red leaves with good bolt resistance. Ready in 45 days. Widely available.

Lettuce, leaf: Simpson Elite
An improvement of 'Black-Seeded Simpson', this has brighter green leaves and better bolt resistance and tastes better in summer heat than most lettuce. Ready in 48 days. Burpee, Johnny's, Park, Pinetree, Southern Exposure.

Marigold, Signet: Gem mix
All marigolds (*Tagetes* species) are edible, but Signet marigolds (*T. tenuifolia*) have the best flavor. The Gem mix contains seeds of orange, pale yellow, and gold flowers, or you can buy single colors. Pull out petals and add to salads. Cook's Garden, Shepherd's.

Mustard: Mizuna
Harvest young, as a salad green in spring or fall. Cut the leaves to 1 inch from the ground

and the plant will regrow. Ready in 35 to 40 days. Cook's Garden, Johnny's, Pinetree.

Nasturtium: Alaska (Tip Top Alaska)
A choice nasturtium selection with green-and-white variegated leaves and flowers in shades of yellow and orange. Leaves and flowers have a spicy flavor. Ready in 60 days. Burpee, Nichols, Pinetree, Shepherd's.

Onion, scallion
Purchase sets in the spring and plant them out to grow scallions. Ready in 30 days. Widely available.

Oregano, Greek
Grow this popular perennial herb to flavor salad dressings. Buy a plant locally, but rub the surface of a leaf and smell it to see if it has a good scent before you buy it.

Parsley
The flat-leaf or plain-leaf types have a richer parsley flavor and are easier to chop. The curled-leaf varieties are more ornamental as garnishes. It is easiest to buy plants at a local nursery.

Pepper, sweet: Bell Boy Hybrid
A standard, productive pepper. Any standard, large bell pepper can be grown in containers. Tie the plants to a bamboo stake to protect them from damage in wind or heavy rain. Ready in 62 to 70 days. Widely available.

Pepper, sweet: Jingle Bells Hybrid
Cute miniature bell peppers form on plants that are 20 inches tall. They taste best when they have turned red, which they will begin to do about 70 days from transplanting. Nichols.

Pepper, sweet: Sweet Pickle
Most small, pointed peppers are hot, but these are sweet. A plant may have yellow, orange, red, and purple fruit all at the same time. Eat them fresh or pickle them. Plants are 12 to 15 inches tall. Ready in 65 days. Park.

Radish: Cherry Belle
A round, red, globe-shaped radish. Ready in 22 days. Widely available.

Radish: French Breakfast
These radishes are somewhat elongated, red at the top, white at the tip. Ready in 25 days. Widely available.

Sage, garden
Culinary garden sage is available in several different leaf colors and forms. All are attractive perennial plants, with edible flowers in spring. Purchase a plant.

Strawberry, Alpine: Ruegen Improved
Pluck small, intensely flavorful strawberries from attractive container plants all summer long. These reproduce by seed or divisions rather than by runner. Buy plants locally if they are available. Perennial plants bear late the first spring from a late-winter indoor sowing, and are hardy to zone 6. Burpee and Nichols carry seed, and Johnny's carries the very similar 'Alexandria'.

Thyme, garden
Culinary garden thyme is a perennial herb available in several different leaf colors and forms. Purchase a plant.

Tomato, large: Better Bush Improved VFF Hybrid
One of the newer long-bearing types that need no staking and have a tidy appearance. The red fruit is 3 to 4 inches across, meaty, and flavorful. Ready in 72 days. Park.

Tomato, large: Husky Gold VF Hybrid
Golden tomatoes and dark green leaves on long-bearing plants. Its short stature makes this variety a good choice for a patio or containers. The fruits average 8 ounces each and have a mild, sweet flavor. Ready in 70 days. Nichols, Tomato Growers Supply.

Tomato, small: Floragold Basket
A choice hanging-basket tomato variety, this makes plenty of yellow cherry-sized fruits. Plant three plants in a 12-inch pot. Ready in 85 days from seed. Tomato Growers Supply.

Tomato, small: Toy Boy VF Hybrid
These 14-inch-tall plants, which produce 1½-inch red fruits, can cascade from hanging baskets, four to a 12-inch container. Ready in 58 days. Tomato Growers Supply.

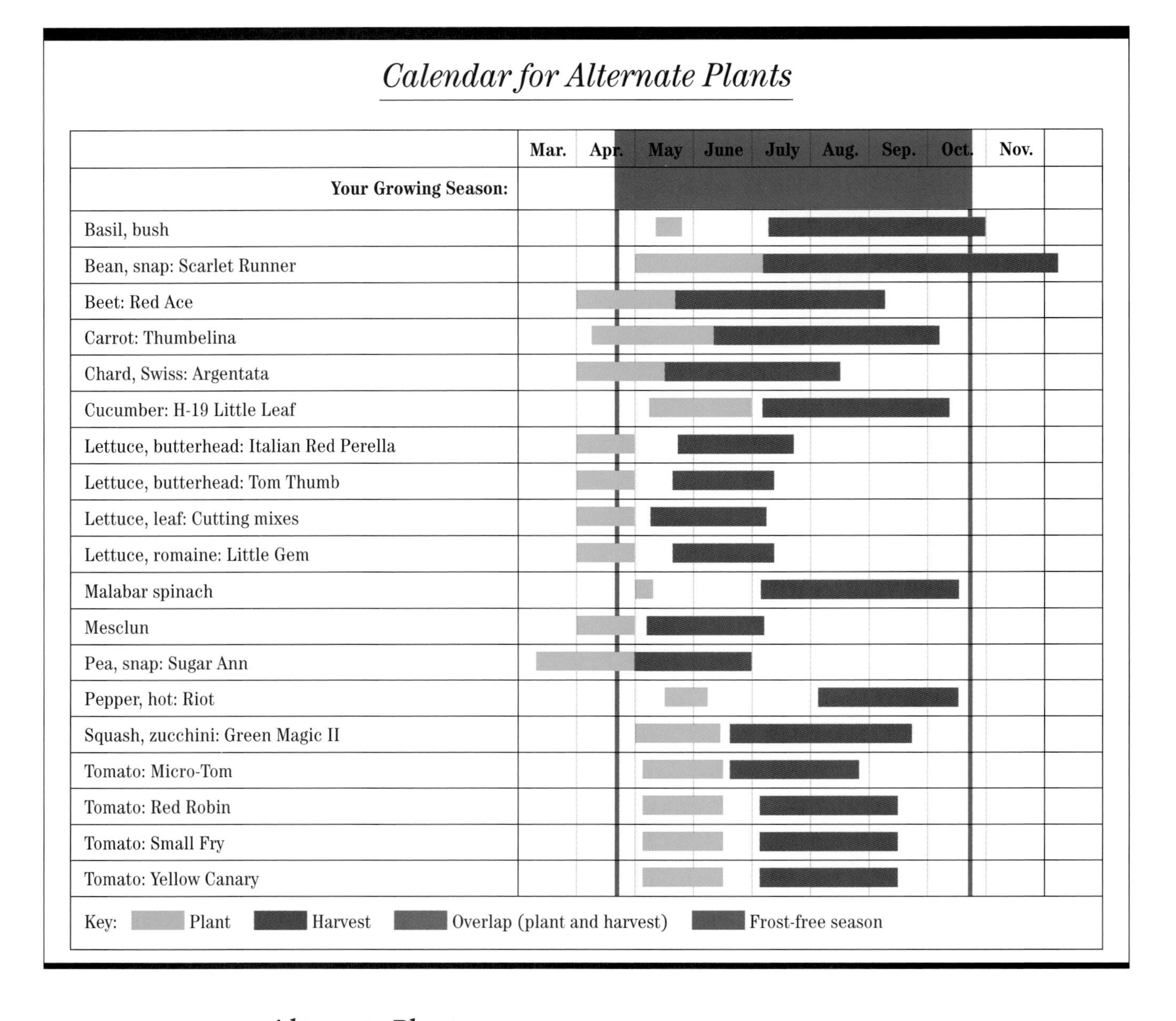

Alternate Plants

This alternate list consists of crops and cultivars not in the actual gardens, but particularly suitable for container gardens. Select from here if you wish to substitute.

Basil, bush (globe basil)
This neatly rounded plant, only 8 inches tall, has small leaves with a delicate, spicy flavor. It is attractive in a container. Widely available.

Bean, pole, snap: Scarlet Runner
Its beauty and high productivity make this bean a good choice for container plantings. Does well in cool summers. Ready in 70 days. Widely available.

Beet: Red Ace
These beets develop fast and stay tender and sweet for a long time. They have good disease tolerance. Ready in 53 days. Field's, Park, Pinetree, Territorial, Vermont Bean Seed.

Carrot: Thumbelina
A ball-shaped carrot that can be grown in a 6-inch-deep container or in hard clay garden soil. 'Parmex' and 'Planet' are similar. Ready in 60 to 70 days. Widely available.

Chard, Swiss: Argentata
An heirloom Italian variety with a particularly mild, sweet flavor, this has silvery white stems and crinkled dark green leaves. Ready in 55 days. Shepherd's.

Cucumber: H-19 Little Leaf
Many qualities make this compact cucumber desirable for container growing. It has small, attractive leaves and many branches that will trail prettily or climb a trellis. The medium-length fruits are good fresh or pickled. And the plants are tolerant to seven common diseases. Ready in 55 days. Johnny's.

Lettuce, butterhead: Italian Red Perella
This baby Bibb-type lettuce is 6 to 7 inches across, green at the base and cranberry red at the tips. It is an Italian heirloom. Ready in 52 days. Shepherd's.

Lettuce, butterhead: Tom Thumb
An old favorite in English gardens, 'Tom Thumb' is a 4- to 6-inch light green Bibb-type lettuce. Ready in 55 days. Widely available.

Lettuce, leaf: Cutting mixes
These carefully chosen mixes of several lettuce varieties are meant to be scatter-sown and harvested as needed by cutting sections ½ to 1 inch from the ground. Ready in 35 to 45 days. See pages 27 and 28 for more on cutting lettuce and mesclun. Cook's Garden, Park, Shepherd's.

Lettuce, romaine: Little Gem
A baby romaine lettuce, 'Little Gem' makes 5-inch heads. Ready in 50 days. Park, Shepherd's.

Malabar spinach
An attractive climbing plant that grows very large where summers are hot. The thick, bright green leaves can be used raw or can be cooked and eaten like spinach. When eaten raw, the leaves have a somewhat mucilaginous texture. Needs plenty of water. Ready in 60 days. Nichols, Pinetree.

Malabar spinach, red
A version of Malabar spinach that has red leaves or just red veins, but is otherwise the same. Park.

Mesclun
Various blends of strong-flavored salad greens are sold under this name. They are intended to be scatter-sown and then harvested as needed; cut sections of the greens ½ to 1 inch from the ground. Ready in 35 to 45 days. See pages 27 and 28 for more information about mesclun. Cook's Garden, Nichols, Territorial, Shepherd's.

Pea, snap: Sugar Ann
Snap peas are a better choice than shelling peas for a container garden because they are more useful in small quantities. This bush variety is about 18 inches tall and needs no staking. It bears 2½-inch-long pods after 52 days. Abundant Life, Gurney's, Johnny's, Nichols, Pinetree, Territorial.

Pepper, hot: Riot
These mounded 10- to 12-inch plants are covered with spiky little hot peppers. They ripen to red 75 to 90 days after transplanting. Territorial.

Squash, zucchini: Green Magic II
A compact bush zucchini. Stocky, thick-stemmed, 18-inch plants have solid dark green fruits. Ready in 48 days. Park.

Tomato, small: Micro-Tom
Sold as the world's smallest tomato variety, 'Micro-Tom' grows 5 to 8 inches tall in a 4-inch pot, with small fruits. Good for patios, windowsills, garden borders. Ready in 85 days from seed. Tomato Growers Supply.

Tomato, small: Red Robin
Really good taste in a dwarf tomato variety that can be grown in a 7-inch pot. Red fruits are 1¼ inches in diameter. Worth trying indoors. Similar plant and fruit size to 'Yellow Canary'. Ready in 63 days. Park, Tomato Growers Supply.

Tomato, small: Small Fry VFN Hybrid
The red fruits are the 1-inch cherry type, which grow in clusters of 7 to 8. The plant forms compact vines with a long fruiting period. Ready in 65 days. Tomato Growers Supply.

Tomato, small: Yellow Canary Hybrid
This tomato, a dwarf for containers, grows well in a 7-inch pot. The yellow fruit is cherry sized. Ready in 63 days. Park, Tomato Growers Supply.

A "GEE WHIZ" GARDEN

This is a garden of out-of-the ordinary vegetables—the ones that are likely to inspire a "Gee whiz, that's different!" or even a "Wow, I never saw anything like that before!" They are fun to grow, fun to show to friends, and fun to use. These are, in other words, perfectly good, edible crops (except for the ornamental gourds) that are also surprising in one or more ways.

About the Plan

Some of the vegetables in this garden are not the color you expect them to be. Purple, blue, and red are alternate colors for a number of common vegetables. In some crops, such as cabbage, potatoes, and okra, the color remains when the vegetable is cooked. In others, including snap beans, pepper, and broccoli, it disappears during cooking, leaving a bright green vegetable. Both kinds of purple coloring are beautiful in the garden. The pole snap bean 'Trionfo Violetto' is an especially beautiful variety with purple pods and flowers and purple-tinged leaves.

In the case of eggplant, we are used to seeing fruit that is a deep, rich purple. In this case

Gee Whiz Garden Plan

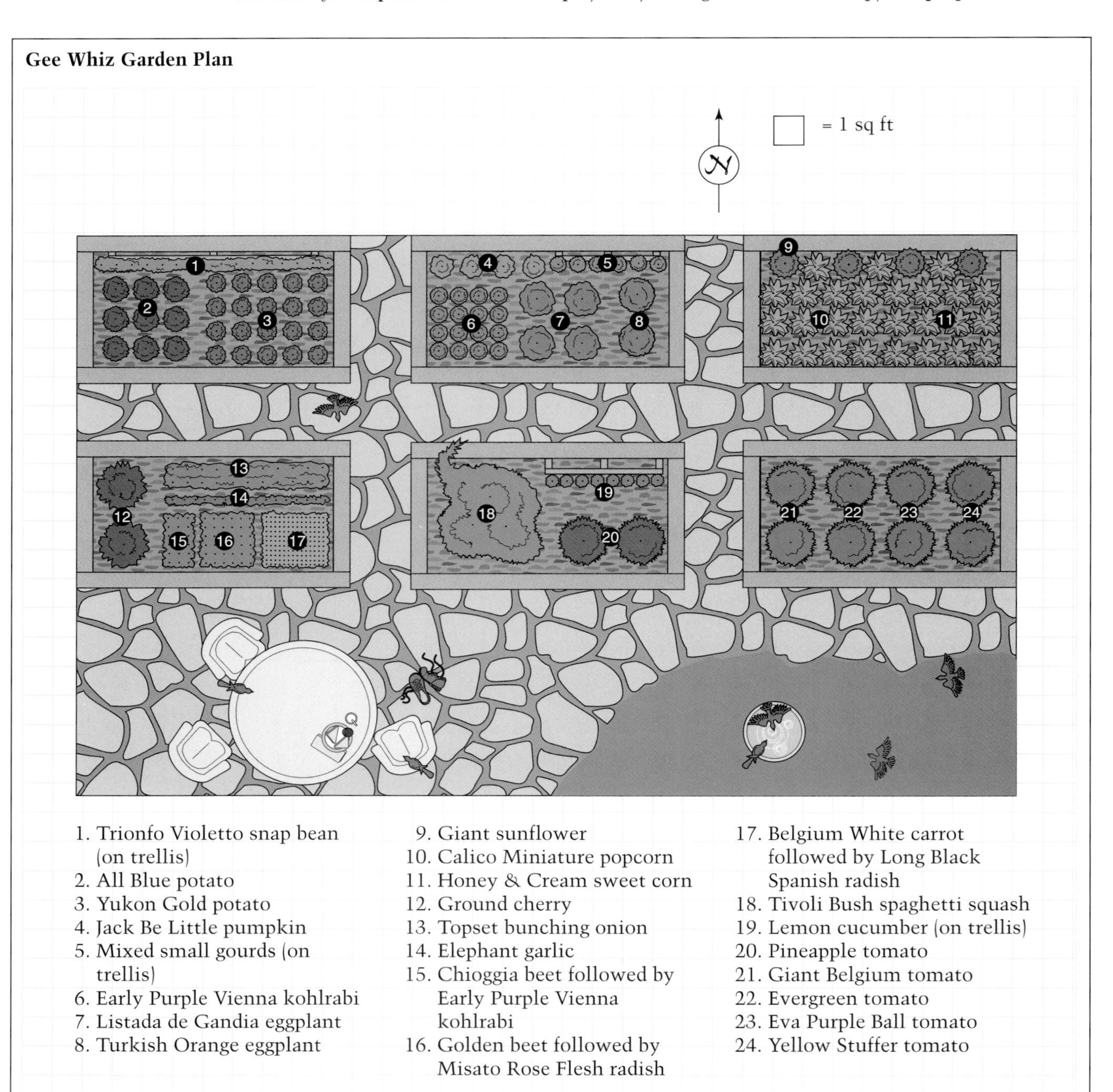

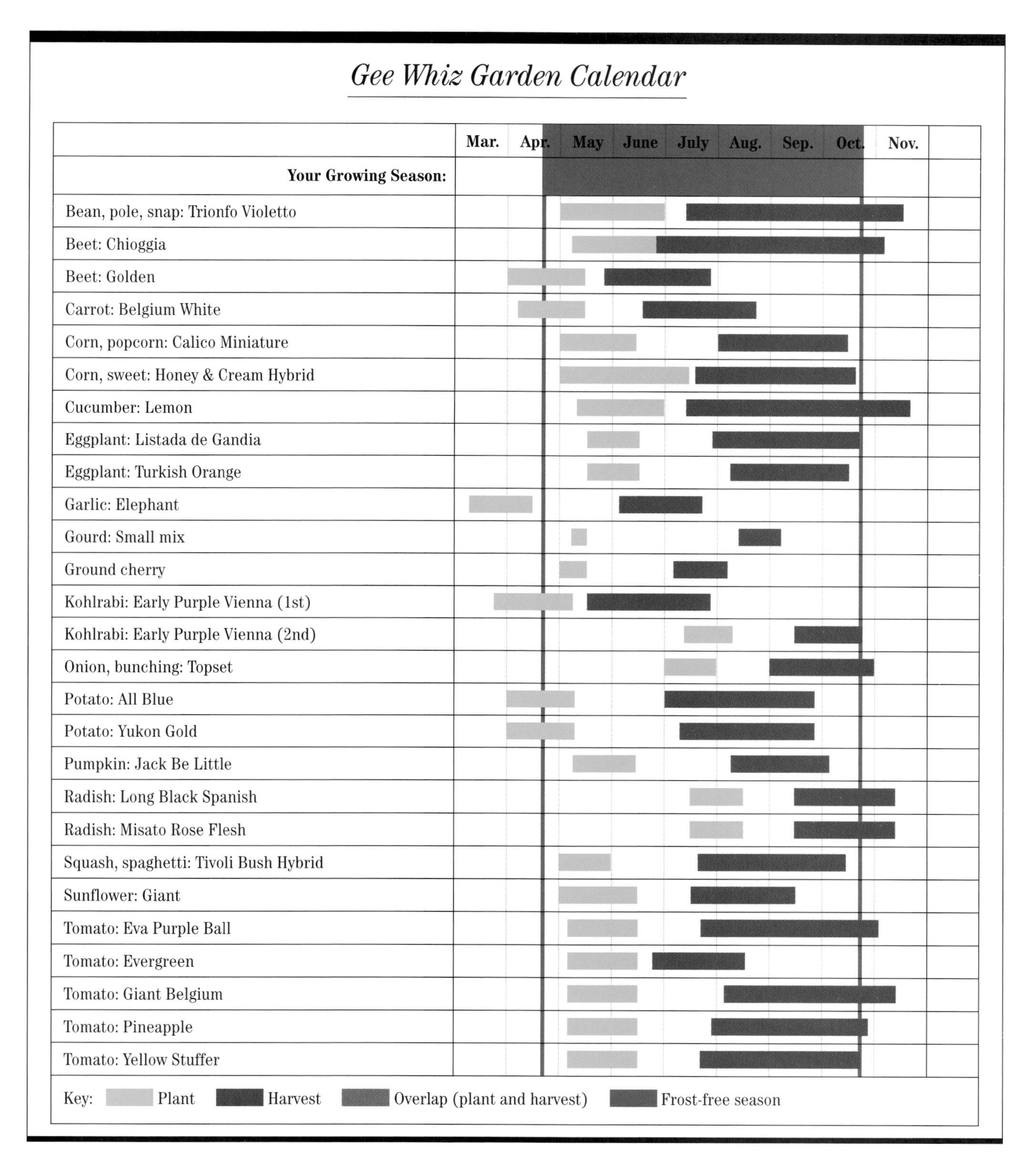

what is surprising is to see other colors, such as a marbled lavender and white, all white, or orange, and to discover that they taste good too! Or consider the carrot. 'Belgium White' is a mild, sweet carrot that is white rather than the familiar orange color.

Tomatoes also come in some surprising colors. You have probably seen yellow or even orange-fruited ones, but how about purple, pink, or yellow-and-red striped? And you have heard of fried green tomatoes, but there is also a tomato called 'Evergreen' that is green when it is ripe.

Really Big Vegetable Varieties

Perhaps you have heard of pumpkin weigh-offs, with the winning fruit weighing in the hundreds of pounds. Or sunflower contests, with winning seedheads exceeding 20 inches in diameter. Gardeners often compete to see who can grow the largest specimens of several different crops, and some gardeners just try to grow the big ones for the fun of it, contest or no. Following is a list of some potentially giant varieties.

If you plan to enter a competition, it is best if you can obtain seed of varieties or strains that have been proven to have the potential to become huge. Be aware that these plants may become quite large in most cases, but some will do so only if they are in a very favorable environment, happy with the climate, soil, fertility, and watering regimen. In some cases, winners go to great trouble to ensure that conditions are perfect. And in some cases, your conditions will not support the particular crop well enough to make a giant, period. For example, short, cool summers will not make the biggest tomatoes, and gardeners with clay soil are not likely to be able to grow record-breaking radishes. Nevertheless, if your conditions are reasonably good, it is great fun to try to grow the big ones.

Garlic: Elephant
See the main listing for description.

Kohlrabi: Gigante
In England, where vegetable size competitions are very popular, this kohlrabi variety set a world record of 62 pounds. Nichols.

Onion: Ailsa Craig Exhibition
The world record onion was a 'Kelsae Sweet Giant' that weighed 7 pounds 7 ounces. Pinetree Garden Seeds reports that 'Ailsa Craig Exhibition' is, on the average, a larger onion, with the potential to become huge. 'Ailsa Craig' is a sweet white onion. Ready in 105 days. Johnny's, Pinetree, Vermont Bean Seed.

Pepper, sweet: Big Bertha Hybrid
These green bell peppers grow 8 inches long by 4 inches wide, and often weigh more than a pound each. They ripen to red. The tobacco mosaic–resistant plants are 24 to 26 inches tall. Ready in 75 days. Field's, Gurney's, Vermont Bean Seed.

Pumpkin: Dill's Atlantic Giant (Atlantic Giant PVP)
This is the variety that most often wins contests. The largest fruits have weighed in at more than 700 pounds, but if you top 200 you should consider it a success. Plants that produce the biggest fruits require 100 to 500 square feet and quite a bit of pampering. Ready in 120 days. Field's, Gurney's, Nichols, Pinetree, Southern Exposure.

Radish: Sakurajima Mammoth
Plant seed in midsummer for a mild, sweet, fall and winter radish that, in Japan, often grows to more than 100 pounds. Ready in 70 days. Nichols.

Squash, winter: Blue Squash
This strain of winter squash has grown to more than 600 pounds. It has gray-blue skin and orange flesh. Nichols.

Sunflower: Giant
This strain of sunflower grows 14 feet tall, with a head typically up to 18 inches across. Nichols.

Tomato: Delicious
A fruit of this tomato variety holds the world record for weight: 7 pounds 12 ounces. Most of the fruits will be more than 1 pound, and, who knows, maybe you'll grow a giant one. This is a good-flavored tomato that resists cracking. Ready in 77 days. Burpee, Field's, Gurney's, Southern Exposure, Tomato Growers Supply.

Watermelon: Carolina Cross
A fruit of this watermelon variety has weighed in at 262 pounds. With enough room and good conditions, you will be able to grow at least 200-pound fruits. They are oval, striped, with sweet, fine-grained flesh. Burpee, Field's, Gurney's, Nichols.

The shape may be part of the surprise. 'Yellow Stuffer' tomatoes look like yellow bell peppers, and can be stuffed and baked like peppers. The radish 'Long Black Spanish' is not only strikingly black-skinned, but is up to 10 inches long. Crisp rounds of this black-and-white root are a stunning addition to a salad.

In other cases, the size is the main source of amazement. Some varieties are unusually large, others unusually small. Though you won't have room in a small garden to grow many giant vegetables, a few kinds aren't space hogs. Giant sunflowers, with heads up to 18 inches across, take up little more room than regular 'Mammoth' sunflower plants. Another small-space giant is 'Elephant' garlic. The giant cloves of this crop are the result of a cross made by the plant breeder Luther Burbank between common garlic and a wild garlic from Chile. While it does not keep as well as regular garlic, and so should not be grown as a main storage crop, it has a milder flavor and is thus very useful in cooking. And it certainly will inspire a "Gee whiz" or two!

Many gardeners try to produce large tomatoes, and some varieties are known for their

ability to produce giant fruits. Two such tomatoes are the dramatically striped and marbled 'Pineapple' and the dark pink 'Giant Belgium'. The world champion tomato weighed 7 pounds 12 ounces, although the variety that produced it, 'Delicious', is more likely to have fruit that weighs somewhat less than that. For more information on giant vegetables, see page 74.

Perhaps more suited to a small garden are varieties that are diminutive by comparison with others. The 'Jack Be Little' pumpkin, for example, is charmingly tiny. The plant makes six or more fruits, each weighing only 3 or 4 ounces. 'Jack Be Little' pumpkins are good to eat (they are great for stuffing) and can be used for decoration. The vines are short, and if you want them to take up even less room, you can trellis them.

'Calico Miniature' popcorn is also ornamental and edible. The small ears were selected for kernels in particularly beautiful color combinations. In addition, some ears have bright purple husks. And yes, the corn will pop. To make a miniature trio for fall decoration, grow a mix of small gourds as well. These are not edible, but they complete a charming fall display.

Finally, some crops are surprising because they are new to many gardeners. Spaghetti squash is one of these. Although it is grown just like a winter squash, it is eaten very differently. Inside each fruit is a mass of "vegetable pasta," ready to cook and serve with a tomato sauce.

The ground cherry is an old crop that is rarely grown and rarely if ever available to nongardeners. This tomato relative bears cherry-sized golden yellow fruit in papery husks rather like those of the tomatillo. It has a sweet and tangy flavor that makes it good for snacking or for desserts. When you are giving a friend the grand garden tour, save the ground cherry for last—a taste of this delicious fruit is the perfect finale.

Plant List

Bean, pole, snap: Trionfo Violetto
Dark green leaves with purple veins, deep lavender flowers, and deep purple bean pods make this one of the prettiest plants you can grow in your vegetable garden, and the beans are great eating as well. Pods cook to green. Ready in 62 days. Cook's Garden, Johnny's.

Beet: Chioggia
Alternate bands of white and red make slices of this beet look like targets. The flavor is sweet and mild. Ready in 52 days. Widely available.

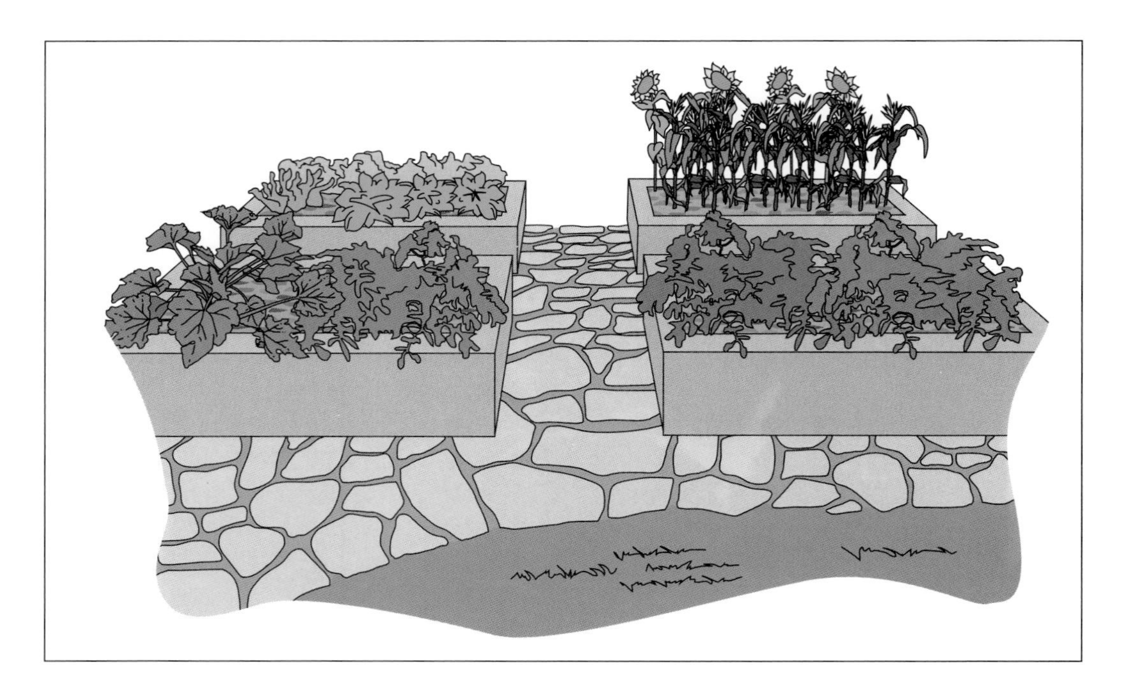

Beet: Golden
These beets are a deep, rich yellow. The germination rate is not as high as for other beet varieties, so sow the seed more thickly than usual. Once up, they are as easy as other beets. Ready in 55 days. Widely available.

Carrot: Belgium White
A carrot of a different color. The white roots are mild flavored, best grown in cool weather. Ready in 75 days. Nichols.

Corn, popcorn: Calico Miniature
The 4-inch ears look like miniature multicolored ornamental corn, but the kernels are popcorn. Some of the husks are purple. Ready in 90 days. Field's. Shepherd's carries a similar 'Calico' popcorn, with 8-inch ears.

Corn, sweet: Honey & Cream Hybrid
These ears combine creamy white kernels with sweet yellow ones, for a taste treat and a visual surprise. Ready in 78 days. Gurney's, Pinetree.

Cucumber: Lemon
These crisp, oval cucumbers really do look like big lemons, though they are better eating at the pale yellow-green stage, before they turn fully yellow. Ready in 64 days. Widely available.

Eggplant: Listada de Gandia
Sure to attract attention, these eggplants are purple with white markings that make them look like marbleized bowling balls. They are 5 to 6 inches long, with white, mild-flavored flesh. Most productive in hot weather and may not flower or set fruit in northern gardens. Ready in 75 days. Southern Exposure.

Eggplant: Turkish Orange
Even more surprising than a white eggplant is this brilliant orange one. Fragrant and sweet-flavored 2- to 4-inch fruits are borne on 18- to 22-inch, insect-resistant bushes. To eat them without peeling, pick just as the fruit develops an orange blush. Ready in 85 days. Abundant Life, Seeds of Change, Southern Exposure.

Garlic: Elephant
A head of elephant garlic, consisting of 8 to 10 giant garlic cloves, weighs up to a pound and a half. The flavor is mild. The cloves do not store as well as regular garlic. Plant in spring and again in midsummer. Widely available.

Gourd: Small mix
Gourds are not an edible crop but make a great addition to fall decorations. These small ones are the easiest to grow. Train them on a 5-foot trellis. Ready in 98 days. Field's, Gurney's, Johnny's, Southern Exposure.

Ground cherry (husk cherry, yellow husk tomato)
A tomato relative that bears golden yellow fruits in papery husks. The flavor is sweet and

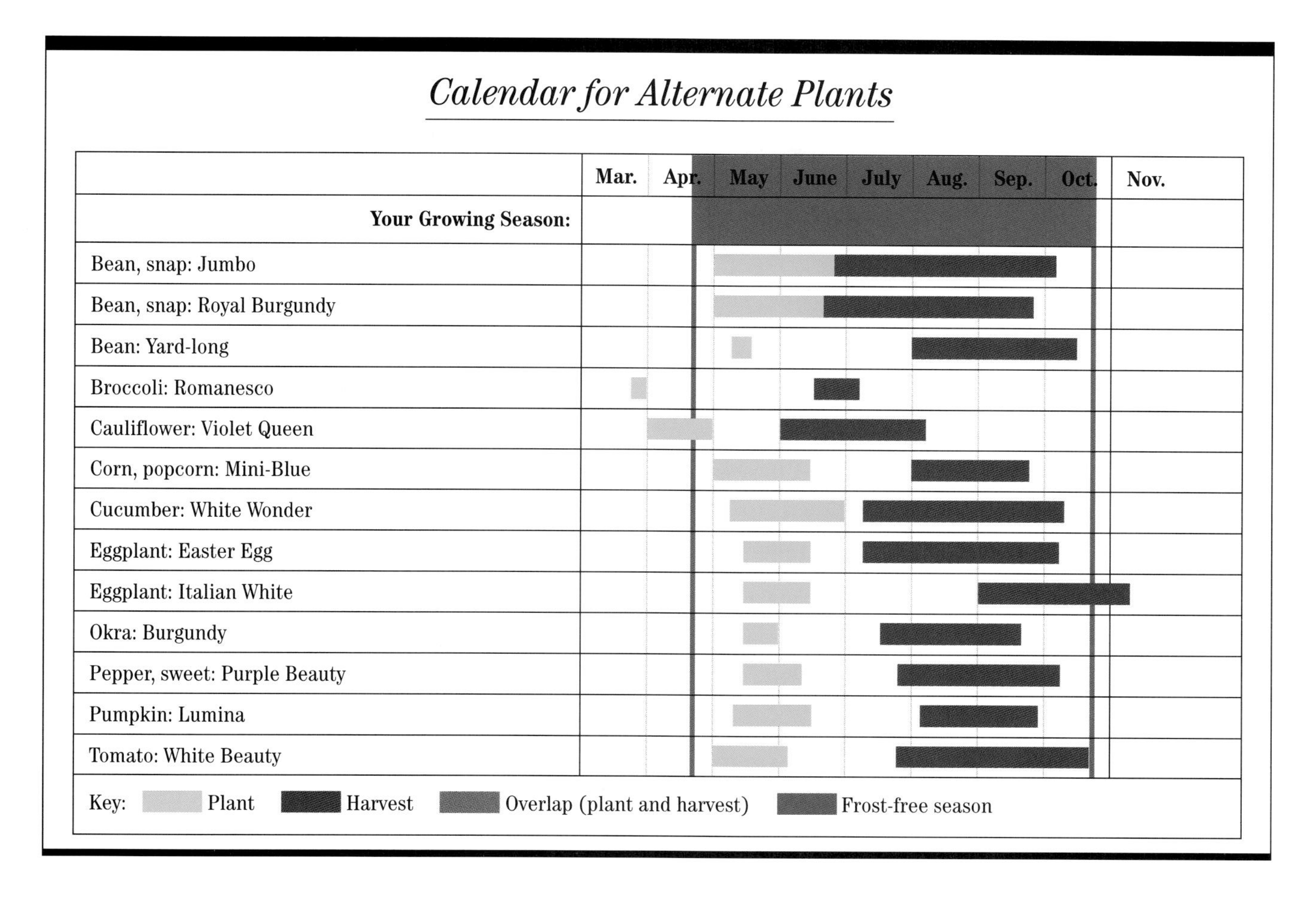

tangy, good for snacking, dessert, or preserves. It thrives in cooler weather than that needed for tomatoes. Plant in spring and again in midsummer. Ready in 90 days. Gurney's, Johnny's, Nichols, Southern Exposure, Territorial.

Kohlrabi: Early Purple Vienna
Kohlrabi always looks like a plant from Mars, with its leaves growing from a big bulblike stem. When it's purple, it looks even stranger, but the stem bulb is white inside, crisp and mild, and good raw or cooked. Plant in spring and again in midsummer. Ready in 60 days. Widely available.

Onion, bunching: Topset (Egyptian, Walking)
This heirloom onion is grown for its production of fine scallions in fall and spring. The plants make reproduction bulblets on top of tall stems. The heads of the bulblets, twisted and spidery, are attention-getting. In midsummer, the bulblets may be cut off and planted for the fall and next spring's crop. Abundant Life, Southern Exposure.

Potato: All Blue
Blue inside and out, but tasty and productive. Imagine blue mashed potatoes! Ready in 110 days. Field's, Gurney's, Territorial.

Potato: Yukon Gold
Both the skin and flesh of this delicious potato are yellow. Plants are virus resistant. Ready in 60 to 90 days. Field's, Gurney's.

Pumpkin: Jack Be Little
True pumpkins that weigh only 3 to 4 ounces each. They are scalloped and flattened in shape. Very ornamental, and good to eat too. Train vines on a 5-foot trellis to save space. Ready in 100 days. Widely available.

Radish: Long Black Spanish
Roots are cylindrical, up to 10 inches long and a bit over an inch in diameter. The skin is a pure, flat black. The flesh is bright white. Ready in 60 days from midsummer plantings. Nichols, Pinetree, Southern Exposure.

Radish: Misato Rose Flesh
A globe-shaped radish that is white-skinned with green shoulders. When you cut it open, the inside is a pretty pink. Up to 4 inches in diameter, 1½ pounds. Ready in 60 days from midsummer plantings. Nichols, Park, Vermont Bean Seed.

Squash, spaghetti: Tivoli Bush Hybrid
A surprising vegetable that many treasure for the pastalike strands it contains. This bush version can be spaced as close as 2 feet. Fruits weigh 2 to 4 pounds. Ready in 98 days. Field's, Gurney's, Pinetree.

Sunflower: Giant
You can expect this strain to produce 18-inch-diameter seedheads on 14-foot plants. Ready in 110 days. Nichols.

Tomato: Eva Purple Ball
These pink-purple tomatoes are 4 to 5 ounces each and have excellent flavor. They resist disease well, and perform well in hot, humid areas. Ready in 78 days. Southern Exposure.

Tomato: Evergreen
Produces medium- to large-sized fruits that are green when fully ripe. They have a mild and delicious flavor. Ready in 72 days. Tomato Growers Supply.

Tomato: Giant Belgium
Meaty, dark pink tomatoes that often weigh in at 3 pounds or more. Sweet and delicious. Ready in 90 days. Tomato Growers Supply.

Tomato: Pineapple
These tomatoes are unusual on two accounts. First, they are mammoth, up to 8 inches in diameter, or 4 pounds. Second, they are bicolored. The yellow skin is splashed with red streaks at the blossom end; the flesh is marbled. It tastes rich and sweet. Ready in 85 days. Tomato Growers Supply.

Tomato: Yellow Stuffer
Tomatoes are closely related to peppers, and you can certainly see that here. These tomatoes look like yellow bells and can be stuffed in the same way. Ready in 76 days. Gurney's, Tomato Growers Supply.

Alternate Plants

Bean, bush, snap: Jumbo
Romano-type snap bean pods, up to a foot long, that resulted from a cross between 'Romano' and 'Kentucky Wonder' plants. The plants are very productive and grow faster in cooler conditions than most snap beans. A good variety for freezing. Ready in 55 days. Gurney's, Nichols, Park, Territorial.

Bean, bush, snap: Royal Burgundy
The pods, a striking deep purple, just keep coming on these long-bearing bushes. Flowers are medium purple. The pods turn green after about 2 minutes at boiling temperatures, giving a good blanching indicator for freezing. Ready in 51 days. Widely available.

Bean: Yard-long (asparagus bean)
These near relatives of cowpeas are used in Chinese cooking. They need high temperatures to thrive, but then produce pods up to 24 inches long. Best for eating when the pods are 12 to 15 inches long. Ready in 80 days. Evergreen, Gurney's, Nichols, Southern Exposure, Sunrise, Vermont Bean Seed.

Broccoli: Romanesco
Striking chartreuse spires have excellent flavor and texture. Plant in May for fall harvest in cold-winter areas. Ready in 85 to 100 days. Abundant Life, Cook's Garden, Johnny's, Nichols, Pinetree.

Cauliflower: Violet Queen Hybrid
The purple heads cook to green, but are handsome in the garden. Some classify this easy-to-grow plant as a purple broccoli, as it makes side sprouts after the central head is cut. However, unless winters are mild, these may form too late to be useful. Ready in 54 to 70 days. Cook's Garden, Johnny's, Nichols, Vermont Bean Seed.

Corn, popcorn: Mini-Blue
Cute 2- to 4-inch ears of unusual deep-blue popping corn. Popped kernels are tender and white. Ready in 110 to 115 days. Gurney's.

Cucumber: White Wonder
A surprisingly white-skinned cucumber, 'White Wonder' is good sliced or pickled. Fruit is 5 to

7 inches long. These plants resist heat well, but are best planted early to avoid late-season diseases. Ready in 58 days. Gurney's, Southern Exposure.

Eggplant: Easter Egg Hybrid
These look just like oversized white eggs, but have a good eggplant flavor. Each 24-inch-tall plant will bear about a dozen fruits. They eventually ripen to yellow but should be eaten while still white. Ready in 52 days. Pinetree.

Eggplant: Italian White
These fruits are round and white, and delicious when picked at baseball size. They are early, mild flavored (never bitter), and productive. Ready in 78 days. Seeds of Change.

Okra: Burgundy Hybrid
A beautiful plant with burgundy-colored branches and stems, burgundy-veined leaves, and pale yellow blossoms. The fruits are red, keep their color after cooking, and are tender almost until they reach their mature 7½-inch length. Plants vary from 2½ feet tall in northern gardens to 4 feet in the south. Ready in 60 days. Field's, Johnny's, Nichols, Southern Exposure, Vermont Bean Seed.

Pepper, sweet: Purple Beauty
These beautiful peppers are dark purple, ripening to red. At the purple stage they are firm and tart. For purple color, use raw, as they cook to green. Ready in 70 days. Cook's Garden, Nichols, Pinetree, Territorial.

Pumpkin: Lumina
A ghostly white pumpkin that makes an unusual ornament or jack-o'-lantern, this is also thick-fleshed and flavorful. The skin may take on a blue or green cast if plants are stressed. Susceptible to squash bugs. Ready in 85 days. Burpee, Cook's Garden, Field's, Park, Pinetree.

Tomato: White Beauty
A mild, sweet tomato that is creamy white inside and out. The flesh is meaty, with few seeds. The fruits average 8 ounces. Ready in 85 days. Tomato Growers Supply.

Vegetable Gardening Tips

In addition to having an easy vegetable garden, you want to have a successful one. Follow the pointers in this chapter for a healthy, bountiful, and enjoyable garden.

Like most activities, gardening is easiest if you know what you are doing and do it right. This chapter helps beginners to do it right the first time. Experienced gardeners might also learn a few tricks for easier gardening.

There are many ways to approach gardening. Each method works well for some people and in some areas. The methods described here are easy ones that are sure to work for all people in all areas. They'll give you a successful garden that takes a minimum amount of time and effort to maintain.

Some of the tips here are basic gardening good sense, and are known by most experienced gardeners. Others are not so commonly known, but are helpful in saving time and effort. To learn more basic information, ask for vegetable gardening pamphlets at your county Cooperative Extension office. In addition, talk to the Master Gardeners there; they will be able to give you general help, and are also a valuable source of local gardening information, such as which varieties do well in your area, or what pests and diseases you need to watch out for and how to protect against them.

Garden books and magazines are also treasuries of gardening tips. Subscribe to a garden magazine and collect as many gardening books as you have time to read. As you learn more, gardening will become less effort and more fun.

This garden embodies several of the tips presented in this chapter. The beds are small enough that the gardener can easily work in them without walking on the soil. They are raised, which brings them to a more comfortable working height and also creates fast drainage. Intensively planted, they are highly productive. Note too that the paths are covered with bark, eliminating the need for weeding.

SELECTING A SITE

The sun's light provides plants with energy for growth. Without enough light they grow slowly, produce sparsely, and are of poor quality. At a minimum, most vegetables need six hours of full sun per day and bright light the rest of the day. If the garden has a shadier end, choose vegetables for that area from the list opposite.

If you have a choice, locate the garden on a south-facing slope, where the soil will warm up faster in the spring and stay warm longer in the fall than will the soil on a north slope, on top of a hill, or in a valley.

Most vegetables must have good drainage. If your soil drains slowly (if puddles stand for several hours after a rainfall), select a sloping site, or garden in raised beds—raising a bed a few inches and incorporating plenty of organic matter usually improves drainage.

A sod cutter can be rented from most rental agencies. It cuts the sod into strips, slicing it loose from the soil below so that it can be rolled up and removed.

PREPARING THE SOIL

The soil in a vegetable garden should be as rich and friable as possible. The best way to improve any garden soil is to dig in large amounts of organic matter before you garden in that spot for the first time. Thereafter, mulch each year. Some organic matter, such as compost and manure, contains plant nutrients and serves as an initial fertilizing (although plants will benefit from one or two more feedings as they grow). As the organic matter ages, it slowly turns to humus, a dark, sticky material that improves the friability of the soil. After two or three years, the soil will have the color and texture of chocolate cake crumbs.

Breaking Ground

If something is growing on the plot of land you've selected for the vegetable garden, first remove it. Unless the plot is very small, remove turf grass with a rented sod cutter. For a small area, remove the sod with a sharp mattock, cutting the sod at root level.

Don't turn the sod into the soil; it will interfere with cultivation and planting, and tufts of grass will sprout as weeds among your vegetables. Instead, make compost for next year's garden out of the cut turf. Cut it into manageable sections and pile it, grass side down and a foot or two high, in an out-of-the-way place. If it is kept moist, the pile will decay into a crumbly compost.

If the site has loose soil, you can turn it over with a spade, spading fork, or round-pointed shovel. First spread the amendments over the soil, then turn the soil over, mixing the amendments into the top 6 to 8 inches. If the soil is hard and compact, or if the plot is too large to turn over by hand, rent a rotary tiller. Choose the largest one available—the weight and power are necessary to dig into hard soil.

Amending the Soil

Good compost is the best soil amendment; aged (but not weathered) horse or cow manure is probably the next best. If you use some other soil amendment—such as leaf mold, sawdust, chopped straw, or ground corncobs—add a general-purpose fertilizer to it, to provide adequate nutrients. Peat moss is an excellent but expensive amendment. With a little research and ingenuity, you can often locate a supply of free or inexpensive amendment material.

Vegetables That Tolerate Some Shade

Although all vegetables grow better in the sun than in the shade, these tolerate more shade than others.

Arugula	Leek
Beets	Lettuce
Broccoli	Parsnip
Cabbage	Potato
Cabbage, Chinese	Radish
Celeriac	Rhubarb
Chicory	Salsify
Cilantro	Spinach
Cress	Swiss Chard
Endive	Turnip
Kale	

Raised Beds

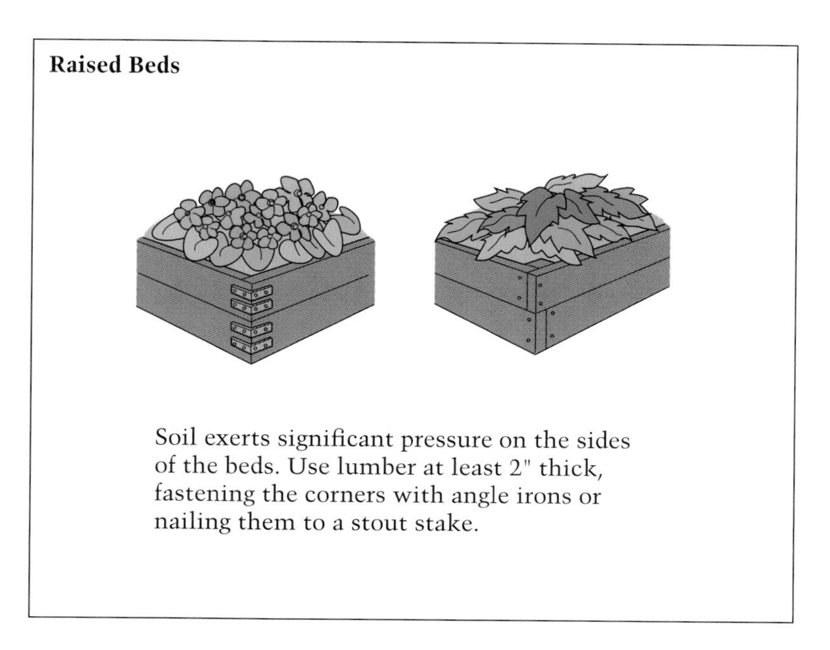

Soil exerts significant pressure on the sides of the beds. Use lumber at least 2" thick, fastening the corners with angle irons or nailing them to a stout stake.

Organic matter buffers acidity and alkalinity, modulating their effects on plant growth, so if you have amended your soil with some organic matter, it should not need any further adjustment. However, if your soil is very acid, add ground limestone when you till the soil initially, and sprinkle some more on the surface each year. If it is very alkaline, add soil sulfur. Your local garden center can tell you about the soil in your area and recommend amounts of lime or sulfur to add.

Shaping the Beds

After you have broken the ground, shape it by shoveling and raking the soil into beds, leaving paths between the beds. Make the beds about 4 inches higher than the paths. This defines the beds and helps the soil drain faster.

You may want to build permanent frames to enclose your garden beds. The frames make the raised beds easier to maintain. If your soil is sandy, frames made from 2×4 lumber will be adequate. Clay soil, which is dense, will benefit from soil amendment and may need 2×6 or even 2×12 lumber frames. These raised beds will improve the aeration of the soil, helping plants grow better in it.

Framed raised beds also allow better access for those who have difficulty stooping. You can construct them with a cap, a board mounted flat on the top of the frame, to serve as a seat. Raised higher, framed beds allow the wheelchair-bound to garden.

Raised beds with permanent frames also make it easier to weedproof the paths. Spread mulch on the paths; the frames will keep the soil and mulch separated. Otherwise, pave the paths with anything that's convenient. If you use pavers, such as bricks or tile, spread weed-blocking fabric under them to prevent weeds from sprouting in the cracks. Old carpet, thick pads of newspaper, discarded boards, and many other materials make good garden paths. Cover unsightly materials with a layer of an attractive mulch. You can leave strips of grass as paths, turning over only the soil in the beds. Leave the paths a convenient width for mowing.

PLANTING

Some vegetables can be sown into the garden as seed, and others are best as transplants. It's simpler to plant seeds directly where they will grow, but starting seeds indoors gets a jump on the season. It takes most vegetable seedlings four to six weeks to grow to transplant size. If they are started in the house six weeks before the weather is warm enough for planting, they will be ready to harvest six weeks earlier, too. In regions with short growing seasons, the only way gardeners can grow tomatoes, peppers, or eggplants is to start them indoors.

Another reason to use transplants is that they are available in nurseries. Using transplants saves some time over planting seed, but has two disadvantages. One is the cost, which is higher than that of seed. The other is that the selection is limited. However, if you need

Top: Plant larger seeds in drills formed by dragging the corner of a hoe. Use a board as a straightedge for straight rows. (Note the corners of this raised bed; these plastic units, available from garden supply stores or catalogs, make the construction of the boxes easy.)
Bottom: These raised beds have no retaining walls, but have been formed by shoveling and raking soil from the paths onto the beds. Slant the sides as shown, to make them more stable.

only two tomato plants, and the variety available is adequate for your needs, nursery plants might be the best buy.

Planting Seeds Outdoors

The first step in planting seeds is to prepare a seedbed. Work the surface until the top couple of inches are fine grained and even. This works best if the soil is just moist when you work it—not too wet and not too dry. Work the surface first with a four-tined cultivator or hoe, chopping up the larger clods, then with a garden rake, until the soil is the texture of cornmeal.

Plant small seeds, such as lettuce and radish, in a furrow made by pressing the corner of a board or the edge of a yardstick into the soil. The furrow should be about ½ inch deep. Place the seeds in the furrow one at a time, spacing them according to the instructions on the seed packet. Cover the seeds with ½ inch of sand. The sand will keep them from drying out without crusting or blocking their emergence, and it will show up well against the soil, marking the row.

Larger seeds, such as those of corn and melons, must be planted deeper. Make a furrow about 2 inches deep with the corner of a hoe, and drop the seeds in the furrow. Use the back of a rake to drag soil back over the furrow, covering the seeds. If the seeds are not to be planted close together, poke individual holes with your finger or a pointed stick, drop a seed into each, and firm the soil to close the holes.

Water well after sowing the seeds. Use the gentlest possible spray to avoid washing small seeds out of the furrow or burying them under soil. If the weather is very hot and dry, shade the seed rows; old boards, burlap, or straw work well. Remove the cover as soon as the seeds begin to break the soil surface.

Planting Seeds Indoors

Seeds are most easily started indoors in peat pots. Made of compressed peat moss, these can be planted in the garden, where they decompose. Fill the pots with purchased seed-starting mix and place them on a watertight tray, such as a cookie sheet. Place two seeds in each pot and cover with ½ inch of potting mix. Label the pots, especially if you are growing more than one variety of a vegetable.

Water the pots thoroughly, let them drain, and then remove any drained water from the tray. Cover the pots and tray with plastic and place them in a warm spot, such as the top of a refrigerator or water heater. They won't need light until the seeds are up. Check them daily, and water if necessary.

As soon as seedlings break the soil surface, move the tray of pots to a cool (60° to 70° F), sunny location. Snip off the weaker of the two seedlings with a pair of scissors.

Whenever the surface of the soil begins to dry, add water to the tray and let the pots soak it up. Once a week, add liquid or soluble fertilizer to the water at one fourth the rate recommended on the fertilizer package.

The seedlings will be ready to plant out four to six weeks after the seeds are sown.

Top: To save space when starting seedlings, plant the seeds close together in a tray for germination. Water the tray well, close it in a plastic bag, and place it in a warm place, such as the top of the refrigerator. As soon as the seedlings begin to appear, move it to a bright, cool location out of direct sun. Bottom: When the second pair of leaves begins to form, gently transplant the seedlings to peat pots or larger containers.

Mulching

If you have planted seeds directly in the garden, spread about 3 inches of mulch over the garden beds after planting, leaving a bare strip 3 inches wide over the seed rows. Once the seedlings are large enough to hold their leaves above the mulch, pull it up to their base.

If you have started your seeds indoors or are using purchased plants, spread the mulch after you have set out the transplants. Spread it about 3 inches thick, to within an inch of the transplant stems. After a week or so, pull the mulch up to the bases of the transplants.

The mulch might be the same material you used as a soil amendment. If it contains plant nutrients, as compost and manure do, it will feed the vegetables as well as shelter the soil. Peat moss is a poor mulch; when dry, it is difficult to wet again, and it can be blown away.

Keep a pile or bag of mulch handy near the garden. If any weeds appear, pull them out and place a little more mulch in that spot. Organic mulches decompose where they are in contact with the soil, so they slowly disappear during the summer and must be renewed periodically.

Some gardeners mulch with black polyethylene (black plastic) or with weed-blocking fabric. Both perform well, stopping weed growth and reducing water loss. However, these materials are more trouble to lay down, less attractive than organic mulch, and inconvenient if you want to replant often, plus they require that you fertilize more frequently.

To mulch with either of these synthetic materials, lay them down after your soil is amended and prepared for planting, but before you plant. Weight the edges with soil to keep them from blowing, or, if you have framed raised beds, cut the fabric with a few inches of overlap on all sides, then tuck the overlap between the soil and the frame with a spade.

Cut Xs where you wish to place transplants and fold back the flaps to plant. Then lay the flaps back against the plant stem. To plant seeds, cut a series of long slits, leaving small "bridges" of uncut film every couple of feet.

WATERING

To be of the best quality, vegetables should never be allowed to suffer drought. In most climates, they will need from 1 to 2 inches of water a week. It can be supplied by either rainfall or irrigation.

If you live in a region with dry summers, you will make your gardening easier by installing a drip irrigation system. Kits are available in most garden centers. The kind that releases water through a porous hose is generally the simplest to install and care for. A worthwhile addition to the kit is a simple battery-operated timer, which will automate the process, freeing you from the chore of daily watering.

When using a plastic film mulch, weight down the edges, cut X-shaped openings in the plastic, and plant through them. This white plastic, which reflects more light and heat than black plastic, keeps the soil slightly cooler. It also reflects extra light onto the bottoms of the plant leaves, perhaps helping them to grow faster.

Porous hoses use the drip principle of watering. They leak drops along their whole length, watering slowly without wetting the whole soil surface. This type is made from recycled automobile tires.

Lay the porous hoses on the surface of the soil and place the mulch on top of them. In most soils, the hoses can be as much as 2 feet apart and still wet all the soil between them.

Your garden's need for water changes as the days grow longer or shorter and as the plants grow. Keep a close eye on the garden to be sure all the plants are getting enough water. The timer setting should probably be changed at least twice a month.

In the absence of rain, most gardens need watering every couple of days during the hottest weather, as little as once a week in cool, overcast weather. The garden should be watered whenever the soil a couple of inches below the surface is just moist—not dry and not muddy. After watering, the soil should be wet at least 12 inches below the surface. Check the water's penetration by digging a hole or by using an inexpensive water meter.

Gardeners who water by hand most commonly use sprinkler irrigation. Set the sprinklers to apply water only as fast as the ground can absorb it without runoff. A very long, slow watering generally does this best. Use an oscillating or reciprocating sprinkler head.

For sprinkler systems, there are timers that turn the water off, either after a certain amount of water has passed through the device or after a certain time. With this type of timer, you can set the sprinkler and leave it to turn itself off.

Some mulches interfere with sprinkler irrigation. Absorbent material, such as sawdust, must be thoroughly wet before any water passes through to the soil below. In such cases, instead of a sprinkler use a drip irrigation system under the mulch, or lay a temporary soaker hose on top of the mulch—the water from the soaker hose will reach the soil after wetting the mulch only slightly.

FERTILIZING

Soil amendments and mulches that contain plant nutrients, such as compost and manure, feed the plants as water passes through them and into the soil. With a couple of inches of compost or composted manure on the soil, you won't need to add any additional fertilizer. In dry-summer areas, if you have a drip system, use a sprinkler to irrigate the garden every couple of weeks so that the nutrients in the mulch will be carried to the plant roots.

In gardens without a nutrient mulch, feed plants a couple of times, especially as they begin to get large. The faster they grow, the greater their need for nutrients.

The fertilizing method depends on your watering method. If you are counting on rain or are using sprinkler irrigation, the easiest way

Top: This type of timed-release fertilizer pellet can feed plants for up to a year. Place one or two in each pot as you fill it with soil. Bottom: Bamboo, being light and strong, makes excellent pea supports. These tall posts are lashed to a horizontal rail at the top, then the middle rails are woven between them. In the foreground is an old-fashioned glass cloche—a bell jar that protects tender plants from the weather.

to feed plants is to sprinkle pelleted fertilizer evenly over the surface of the soil or mulch, then water thoroughly. Water dissolves the nutrients and carries them to the plant roots.

If you have a drip system, the easiest way to feed is to purchase a fertilizer injector—a device that attaches to your drip system and slowly feeds dissolved fertilizer into the irrigation water. Every week or two, fill the injector with liquid or soluble fertilizer.

If your plants are in containers and you are watering by hand, time-release fertilizers are very helpful. These pelleted products release nutrients over a specified period, so you can depend on them for an entire season's growth.

SUPPORTING PLANTS

Although almost any vining plant can be allowed to sprawl on the ground, it will use up lots of space, be difficult to harvest, and be more prone to damage, disease, and insect problems. Vining varieties of beans and peas are usually supported. (Bush types don't need support.) Standard indeterminate tomatoes—those that keep growing and bearing all season—need support. Cucumbers and smaller varieties of trailing melons and squash are often grown on the ground but are more satisfactory if supported.

Supporting Peas and Beans

The easiest support for peas or beans is disposable fencing strung on a simple frame.

Make the frame by planting strong posts about 6 feet apart where you will plant peas or beans. The treated pine poles sold to support landscape trees—about 8 feet long, 2 to 3 inches in diameter, and pointed on one end—make good supports. If your soil is heavy clay or rocky, make a hole with a crowbar first, then use a sledgehammer to drive the post into the hole about 18 inches deep.

Use bamboo or other lightweight poles, about 6 feet long, for the stringers (the horizontal poles). To cover the frame inexpensively, use the garden netting sold to keep birds away from fruit. Purchase or cut the netting so it is the height you want the trellis to be. Weave one stringer through the top and one through the bottom of the netting, then tie the stringers to the support poles.

Supporting Tomatoes

Most determinate tomatoes, which ripen all their fruit at once and then stop growing, don't need support. If they threaten to flop on the ground, place wooden crates or some other support between the plants.

Indeterminate tomatoes, which continue to grow and bear fruit all summer, are best if supported. The simplest support is a wire cage, a cylinder of self-supporting fencing that holds the plant upright. Either purchase a cage made for this purpose, or make your own from woven wire fencing or concrete reinforcing mesh. Use fencing that is rigid enough to support a heavy tomato and that has a mesh large enough for your hand to reach through, to pick tomatoes. A good choice is concrete-reinforcing mesh, which is very rigid and will support the heaviest tomato plant. It's made from thick wire; use bolt cutters or lineman's pliers to cut it.

For small tomato plants, make the cages 18 inches in diameter by 3 feet high. For large plants, they'll need to be 3 feet in diameter by 5 feet high. Further support the largest tomato plants (those that produce very large fruit) with a post driven in one or two sides of the cage to keep it from leaning over under the plant's weight.

Supporting Cucumbers and Squash

Cucurbits—the family to which cucumbers, squash, and melons belong—are frequently grown on the ground; some, such as zucchini, are bushes rather than vining plants and never need support. But many are best when grown on a support.

Because the vining cucurbits are heavy, they need a more substantial support than the fencing on which peas and beans are grown. They are often grown on permanent wooden trellises or on a permanent garden fence. Small-fruited cucurbits, such as cucumbers, can grow on such a support with no further aid. Those with medium-sized fruit, however—such as cantaloupes, acorn squash, and small watermelons—need further support for each fruit to keep it from pulling loose from the vine. As each fruit begins to enlarge, make a sling for it out of stretchable fabric, such as old panty hose or T-shirts, tying the sling to the trellis or fence. Those with large fruit can't be grown on a trellis because of the weight of the fruit.

These tomatoes are growing in a cage made of welded wire. The stakes on the sides help support the cage. When selecting wire for a cage, be sure your hand will fit through the mesh to allow you to pick the tomatoes. At the beginning of the season, tuck errant branches back in the cage. Later, after the plant has been trained, you can let them grow out the sides.

Sources of Seeds and Plants

If local garden centers do not carry the varieties of plants or seeds that you want, try some of these mail-order sources.

Abundant Life Seed Foundation
Box 772
Port Townsend, WA 98368
Vegetable, herb, flower, tree, and shrub seeds. Specializes in heirloom and other open-pollinated seeds.

W. Atlee Burpee & Co.
300 Park Avenue
Warminster, PA 18974
Vegetable, herb, and flower seeds. Herb, fruit, and flower plants.

The Cook's Garden
Box 535
Londonderry, VT 05148-0535
Vegetable, herb, and flower seeds. Excellent selection of salad vegetables, and a selection of edible flowers.

Henry Field's Seed & Nursery Co.
415 North Burnett
Shenandoah, IA 51602
Vegetable, herb, and flower seeds. Fruit, flower, shrub, and tree plants.

Gurney's Seed and Nursery Co.
110 Capital Street
Yankton, SD 57079
Vegetable, herb, and flower seeds. Herb, fruit, flower, shrub, and tree plants.

Johnny's Selected Seeds
Foss Hill Road
Albion, ME 04910-9731
Vegetable, herb, and flower seeds. Varieties are tested in the short summers of Maine.

Kitazawa Seed Co.
1111 Chapman Street
San Jose, CA 95126
Vegetables used in Asian cooking.

Native Seeds/SEARCH
2509 North Campbell Ave. #325
Tucson, AZ 85719
Crops grown by Native Americans, especially from the Southwest and Mexico.

Nichols Garden Nursery
1190 North Pacific Highway
Albany, OR 97321-4598
Vegetable, herb, and flower seeds. Herb plants.

George W. Park Seed Co.
Cokesbury Road
Greenwood, SC 29647-0001
Vegetable, herb, and flower seeds. Herb plants.

Pinetree Garden Seeds
Box 300
New Gloucester, ME 04260
Vegetable, herb, and flower seeds. Their list includes many heirlooms, as well as Italian, French, Latin American, Asian, and Native American crops.

Seeds of Change
Box 15700
Santa Fe, NM 87506-5700
Vegetable, herb, and flower seeds. Specializes in heirloom and other open-pollinated seeds.

Shepherd's Garden Seeds
30 Irene Street
Torrington, CT 06790
Vegetable, herb, and flower seeds. Herb plants. Edible flower section. Specializes in European gourmet vegetable and herb varieties.

Southern Exposure Seed Exchange
Box 170
Earlysville, VA 22936
Vegetable and herb seeds. Specializes in heirloom varieties.

Territorial Seed Company
Box 157
Cottage Grove, OR 97424
Vegetable, flower, and herb seeds. Varieties offered are tested in the cool climate of the Pacific Northwest.

Tomato Growers Supply Company
Box 2237
Fort Myers, FL 33902
A wide selection of tomato and pepper seeds.

Vermont Bean Seed Company
Garden Lane
Fairhaven, VT 05743
Vegetable, herb, and flower seeds. A wide selection of heirloom beans.

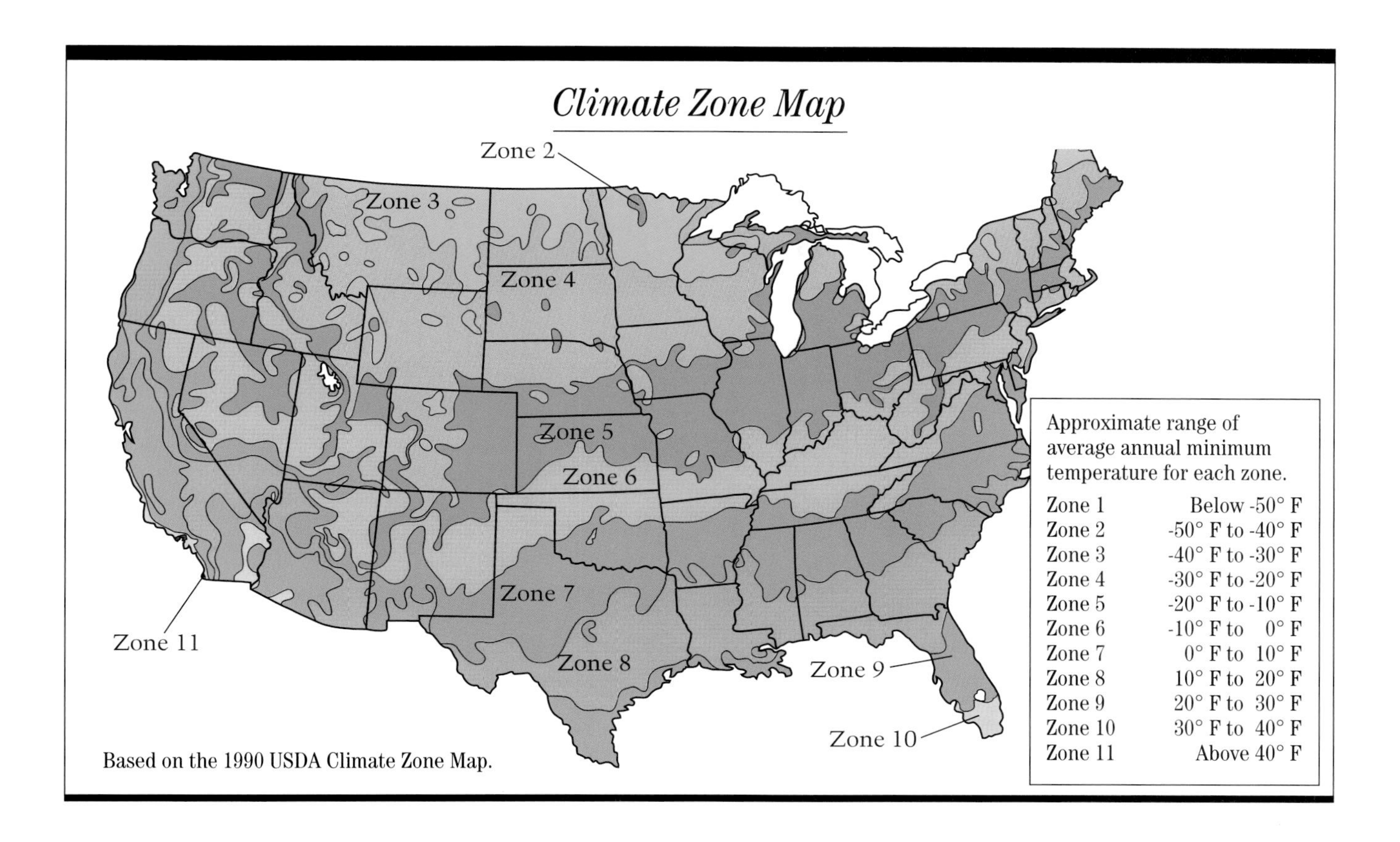

Based on the 1990 USDA Climate Zone Map.

U.S. Measure and Metric Measure Conversion Chart

	Formulas for Exact Measures				Rounded Measures for Quick Reference		
	Symbol	When you know:	Multiply by:	To find:			
Mass (weight)	oz	ounces	28.35	grams	1 oz		= 30 g
	lb	pounds	0.45	kilograms	4 oz		= 115 g
	g	grams	0.035	ounces	8 oz		= 225 g
	kg	kilograms	2.2	pounds	16 oz	= 1 lb	= 450 g
					32 oz	= 2 lb	= 900 g
					36 oz	= 2¼ lb	= 1000 g (1 kg)
Volume	pt	pints	0.47	liters	1 c	= 8 oz	= 250 ml
	qt	quarts	0.95	liters	2 c (1 pt)	= 16 oz	= 500 ml
	gal	gallons	3.785	liters	4 c (1 qt)	= 32 oz	= 1 liter
	ml	milliliters	0.034	fluid ounces	4 qt (1 gal)	= 128 oz	= 3¾ liter
Length	in.	inches	2.54	centimeters	⅜ in.	= 1.0 cm	
	ft	feet	30.48	centimeters	1 in.	= 2.5 cm	
	yd	yards	0.9144	meters	2 in.	= 5.0 cm	
	mi	miles	1.609	kilometers	2½ in.	= 6.5 cm	
	km	kilometers	0.621	miles	12 in. (1 ft)	= 30 cm	
	m	meters	1.094	yards	1 yd	= 90 cm	
	cm	centimeters	0.39	inches	100 ft	= 30 m	
					1 mi	= 1.6 km	
Temperature	° F	Fahrenheit	⁵⁄₉ (after subtracting 32)	Celsius	32° F	= 0° C	
	° C	Celsius	⁹⁄₅ (then add 32)	Fahrenheit	212° F	= 100° C	
Area	in.^{2}	square inches	6.452	square centimeters	1 in.^{2}	= 6.5 cm^2	
	ft^2	square feet	929.0	square centimeters	1 ft^2	= 930 cm^2	
	yd^2	square yards	8361.0	square centimeters	1 yd^2	= 8360 cm^2	
	a.	acres	0.4047	hectares	1 a.	= 4050 m^2	

INDEX

Note: Page numbers in **bold-face** type refer to initial descriptions of varieties; in *italics*, to photographs.

A

Amaranth (hinn choy), 44
  'Red Stripe', 42, **44**
Anise, 51, **54**
  to harvest, 53
Anise hyssop (licorice mint), 27, 51, 57
  described, **54**, 59
Arbor, garden with, 63–64
Artichoke, 61
  to grow, 13
  'Imperial Star', **40**
  Jerusalem. *See* Jerusalem artichoke
  'Violetto', **41**
Arugula (rocket; roquette; rucola), 25, 34, 57, 67, 83
  described, **28**, 36, 59, 64
  flowers, 27
  to grow, 62
Asparagus
  to grow, 13
  'Jersey Giant', 14, **15**
  'UC 157', **18**

B

Balcony, container garden on, 63
  weight limitations of, 62
Bamboo, for supports, *88*, 89
Basil, 27
  'Broadleaf Sweet', **38**
  bush (globe), **70**
  cinnamon, 51, **54**
  globe. *See* basil, bush
  'Purple Ruffles', 25, **28**
  sweet, 51, **54**
  'Sweet Genovese' ('Genovese'; 'Genova Profumatissima'), 25, 34, 57, 67
    described, **28**, 36, 59, 64
  Thai, 42, **44**
Bay (bay laurel), 51, 52, **54**
Beans
  asparagus. *See* beans, yard-long
  bush
    shelling, 'Cannellini' ('Cannelone'), **40**
    snap, 'Roma II', **38–39**
  bush snap
    'Blue Lake 274', 18, **18**
    'Jumbo', **78**
    'Royal Burgundy', **78**
  flowers, 27
  'Genuine Cornfield', **24**
  to grow
    container depth and, 62
    with corn, 48
    support for, 88–89
  lima, 'King of the Garden', 14, **15**
  pole snap
    'Annelino', 34, **40**
    'Blue Lake', **24**
    'Blue Lake Pole', 14, **15**
    'Kentucky Blue', 46, **48**, 57, **60**
    'Kentucky Wonder', **30**, 31, 48, 60, 78
    'Kentucky Wonder Wax', 31, **31**
    'Romano' ('Italian Pole'), 34, **36**, 78
    'Scarlet Runner', *20*, 23, **24**, 63, 70
    'Trionfo Violetto', 63, **64**, 67, 72, **75**
    yard-long (asparagus bean; dow gauk), 42, 44, **44**, **78**
  winged, 42
Beds, garden
  building, 52, 83, *84*
  permanent, 6
  raised, *7*, *80–81*, 82, 83, *83*, *84*
  size of, 6, *80–81*
  terraced, 36
Beet, 62, 83
  'Chioggia', 72, **75–76**
  'Golden', 72, **76**
  'Lutz Winter Keeper', 14, **18**
  'Red Ace', 14, **15**, 25, 28, 61, 70
Borage, 25, 27, **28**, 51, **54**
Box, as container, 63, *63*
Broccoli, 11, 83
  'Calabrese' ('Italian Green Sprouting'), **39**
  'Emperor Hybrid', **33**
  'Minaret', 40
  'Premium Crop Hybrid', 14, **15**, 33, 57, 60
  'Romanesco', **40**, 78
Broccoli raab (rapini), 34, **34**
Brussels sprouts, 11
Burbank, Luther, 74

C

Cabbage, 11, 15, 83
  Chinese (hakusai; pe tsai), 83
  'Michihili', 42, **44**
  see also pak choi
  red, 'Ruby Perfection Hybrid', **33**
  'Salarite Hybrid', 14, **17**, 31
  'Storage No. 4 Hybrid', 18, **18**
Cage, tomato, 89, *89*
Calendar
  Asian garden, *43*
  balcony container garden, *66*, *70*
  basic garden, *16*, *19*
  children's garden, *21*
  deck container garden, *65*, *70*
  "gee whiz" garden, *73*, *77*
  Italian garden, *35*, *38*, *40*, *41*
  Mexican garden, *47*, *50*
  Native American garden, *23*
  ornamental/edible garden, *68*, *70*
  patio garden, *58*
  planting, 11, 13
  salad garden, *26*, *30*, *32*, *33*
  sample, *11*
Cantaloupe, 'Ambrosia Hybrid', 14, **17**, 41, 89
Caraway, 51, **54**
Carrot, 62
  "baby," 64
  'Belgium White', 72, 73, **76**
  'Little Finger', **64**, 67
  'Parmex', 70
  'Planet', 70
  'Rumba', 18, **18**
  'Scarlet Nantes', 14, 25, 57, 67
    described, **17**, 28, 60, 64
  'Thumbelina', *20*, **24**, **70**
Catnip, 51, 53, 54, **54**
Cauliflower, 11, *12*
  'Violet Queen Hybrid', **78**
Celeriac, 83
Chamomile, 27
  German (*Matricaria recutita*), 54
  to grow, 53
  Roman, 51, 52, **54**
  'Treanague', 54
Chan, Peter, 44
Chard. *See* Swiss chard
Chayote, *10*
Chenopodium, spp., 47
  ambrosioides (epazote), 48
  berlandieri, 47
Chervil, 51, **54**
Chicory, 83
  'Catalogna' ('Dentarella'), **41**
Children. *See* garden, for children
Chives, 25, 27, 34, 42, 51, 57, 64, 67
  described, **28**, 36, 54, 60, 65
Chrysanthemum, garland (shungiku), 42, 43, **44**
Cilantro (Chinese parsley; coriander), *10*, 27, 42, 51, 53, 83
  described, **54**
  'Slow Bolt' ('Long Standing'), 42, **45**, 46, 48, 54

Climate zone, map, *91*
Cloche, *88*
Compass, to devise a large, 52
Compost, 82
Containers
 crops for, 63
 fertilizing, 88
 gardening in, 10, 52–53, 56, 62–71
 size of, 62
 types of, 62
 weight of, 62
Coriander
 seeds, to harvest, 53
 see also cilantro
Corncobs, for mulch, 82
Corn (sweet corn), 48, 85
 'Black Aztec', **24**
 'Honey & Cream', 72, **76**
 'Kandy Korn EH Hybrid', 14, **17**, 24, 46, 48
 see also popcorn
Cress, 83
 curled, 54, **65**
 garden, 51, 53, **54**, 64
 upland, 54
Cucumber
 A-frame for, 23
 container depth and, 62
 pickling, 18
 supporting, 63, 72, 88, 89
 'Bianco Lungo di Parigi', **41**
 'H-19 Little Leaf', **71**
 'Lemon', 72, **76**
 'Pickalot Hybrid', 18, **18**
 'Salad Bush Hybrid', **65**, 67
 'Slicemaster Hybrid', 14, **17**
 'Suyo Long', 42, **45**
 'Sweet Success Hybrid', **30**
 'Tasty Green 26 Hybrid', 25, 57, 64, 67
  described, **28–29**, 41, 60, 65
 'White Wonder', **78–79**

D
Daylily (golden threads), 27
"Days to maturity," explained, 11
Deck, container garden for, 63
Depth, planting, 62
Dill, 51, 53, **55**
 'Dukat', 18, **18**, 31, 31
 'Fernleaf', 31, **31**
 seeds, to harvest, 53
Doan gwa. *See* melon, winter
Dow gauk. *See* beans, yard-long
Drainage, 82
Drip irrigation, 86–87
 fertilizing and, 88

E
Edging, plastic, 52
Eggplant
 'Asian Bride', **45**, 66, 67
 'Dusky Hybrid', 14, **17**
 'Easter Egg Hybrid', **41**, 66, 67, 79
 to grow, 11
  colors of, 72–73
  container depth and, 62
 'Italian White', **79**
 'Listada de Gandia', 72, **76**
 'Orient Express Hybrid', **45**
 'Ping Tung', 42, **45**
 'Rosa Bianca', 34, **36**, 57, 60
 'Turkish Orange', 72, **76**
Endive, 83
 Belgian, *10*
 'Green Curled', 28
 'Très Fine', 25, **29**, 30
Epazote (*Chenopodium ambrosioides*, 46, 48, **48**
Escarole, 'Nuvol', **30**, 34, 36–37
Extension, to the basic garden, 15

F
Fabric, weed-blocking, 86
Fennel, *10*, 27, 51, **55**
 Florence, 34, 55
  'Zefa Fino', 34, **37**
 seeds, to harvest, 53
Fertilizer, timed-release, 88, *88*
Fertilizing, 82, 87–88
Flowers
 edible, 26–27, 59
 poisonous (*list*), 27
Frisée. *See* endive
Fruit tree, 46, 48
 miniature, 63, 67

G
Garden
 Asian, 10, 42–45
 basic, 14–19, *14*
 for children, 9, 20–25
 container, 10
 in containers, 62–71
 design, 51–52
 "gee whiz," 10, 72–79
 herb, 10, 51–56
 Italian, 10, 34–41
 Mexican, 10, 46–50
 Native American, 22
 patio, 10, 57–61
 plans, 13
  extensions to, 13
 Pre-Columbian Mexican (*milpas*), 48
 salad, 9, 25–33
 size of, 6
Gardening
 intensive, *4–5*, 44, *80–81*
 principles of easy, 6–7
Garlic, 25, **29**, 39, 42, 51, 55, 66, 67
 to grow, 53
 to harvest, 53–54, 66
 'Elephant', 55, 72, 74, **76**
 'Italian Purple Skin', 55
Garlic chives (gow choy; nira), 27, 42, 44, 51, 57, 67
 container depth and, 62
 described, **45**, 55, 60, 66
Germination, demonstrating, for children, 23–24
Ginger, 42
Golden threads. *See* daylily
Gourd
 bitter. *See* melon, bitter
 small, 72, 75, **76**
 wax. *See* melon, winter
Gow choy. *See* garlic chives
Ground cherry (husk cherry; yellow husk tomato), 72, 75, **76–77**
Ground cover, 52
 to plant, 53
 thyme as, 51, 56

H
Hakusai. *See* cabbage, Chinese
Harvesting, 15
 dates for, 13
 herbs, 53–54
Herb garden, 10, 51–56
Hinn choy. *See* amaranth
Horehound, 51, 53, **55**
Horseradish, *10*, 51, 53, **55**
 to harvest, 53
Hose, porous, 87, *87*
Hot pepper. *See* pepper
Husk cherry. *See* ground cherry
Hyssop, 51, **55**

I, J
Indoors, starting plants, 11
Jerusalem artichoke, *10*
Jicama, *10*
Johnny-jump-up, 25, 27, 57, 59, 62, 64, 67
 described, **29**, 60, 66

K
Kale, 83
Kohlrabi, 21
 'Early Purple Vienna', 72, **77**
 'Early White Vienna', 31, **31**
 'Gigante', **74**

L
Lamb's quarters (*Chenopodium* spp.), 47
  C. berlandieri, 47
  see also quelite
Leaf mold, 82
Leek, 83
Lemon balm, 51, 53, **55**
  lime-scented, 55
Lemongrass, 42
Lemon tree, 48
Lemon verbena, 51, **55**
Lettuce
  Batavian. *See* lettuce, crisphead
  Bibb-type, 71
  butterhead
    'Buttercrunch', 14, 25, 57, 67
      described, **17**, 29, 60, 68
    'Italian Red Perella', 34, **37**, 71
    'Tom Thumb', **71**
  crisphead
    'Nevada', 25, **30**
    'Sierra', 25, **30**
  cutting mix, **30**, 71
    see also mesclun
  to grow, 25, 62, 83, 85
  iceberg, 25
  leaf
    'Lollo Biondo', 34, **37**
    'Lollo Rossa', 34, **37**, 57, 60
    'Red Sails', *20*, 25, 67
      described, **24**, 29, 68
    'Simpson Elite', 14, **17**, 64, 67, 68
  'Mignonette Green', 24
  'Oakleaf', 25, **29**, 57, 60
  'Red Oakleaf' ('Red Salad Bowl'), 31, **31**, 64, 67, 68
  'Red Salad Bowl'. *See* lettuce, 'Red Oakleaf'
  romaine
    'Little Gem', **71**
    'Parris Island Cos', 14, **17**
    'Plato', 25, 34, 46
      described, **29**, 37, 48
Licorice, 51, **55**
  to harvest, 53
Limestone, ground, 83
Lime tree, 48
Lovage, 51, **55**

M
Mail order, for seeds, 13, 90
Malabar spinach, **71**
Manure, 82
Map, climate zone, *91*
Marigold (*Tagetes* spp.), 60, 62, 68
  'Gem Mix', 25, 27, 57, 59, 64, 67
    described, **29**, 60, 68
  Signet (*T. tenuifolia*)
Marjoram (*Origanum marjoranum*), 51, **55**
Matricaria recutita. *See* chamomile, German
Melon, 85, 88, 89
  bitter (bitter gourd; foo gwa), 42–43, 44, **44**
  'Charmel Hybrid', **41**
  'Chimayo', 48
  'Galia', 46
  'Galin Hybrid', **48**
  'Rio Grande Pueblo Mix', 48
  winter (doan gwa; wax gourd), 42, **45**
  see also watermelon
Mesclun, 27–28, **30**, **71**
  see also lettuce, cutting mix
Milpas, 48
Mint, 27, 42
  growing, 53
  licorice. *See* anise hyssop
  peppermint, 51, **56**
  spearmint, 34, 46, 51
    described, **31**, 37, 49, 56
Misticanza. *See* mesclun
Mitsuba (Japanese parsley), 42, 44, **45**
Mizuna. *See* mustard, Mizuna
Mold, for squash, 23
Morning glory, tepee of, 23
Mulch, 82, 86
  on paths, 83
  watering and, 87
  for weed prevention, 23
Mustard
  flowers, 27
  'Green in Snow', 42, **45**
  'Mizuna', 25, 42, 43, 62, 64, 67
    described, **29**, 45, 68–69

N
Nasturtium, 27, 59, 62
  'Alaska' ('Tip Top Alaska'), 31, 57, 64, 67
    described, **32**, 60, 69
  'Tip Top Mix', 60
  'Whirlybird', 57, **60**
Nepeta cataria, 54
Newspapers, for mulch, 23
Nira. *See* garlic chives

O
Okra, 'Burgundy Hybrid', **79**
Onion
  'Ailsa Craig Exhibition', **74**
  bulbing, 'Bermuda White', 31, **32**, 46, 48, 49
  bunching
    (Egyptian onion; walking onion), 'Topset', 72, **77**
    'White Spear', 14, **17**
  'Granex Hybrid', **19**
  'Kelsae Sweet Giant', 74
  large, 74
  'Mambo Hybrid', 18, **19**
  'Red Creole', **19**
  'Red Mac', 31, **32**
  'Sweet Sandwich Hybrid', 18, **19**
  see also scallion
Oregano, 27, 51, **56**, 62
  Greek, 31, 34, 46, 57, 64, 67
    described, **32**, 37, 49, 61, 69
Origanum
  heracleoticum, 56
  marjoranum. *See* marjoram
  onites, 56
  vulgare hirtum, 56

P
Pak choi, 43
  'Mei Qing Choi', 42, **45**
Parsley, 14, 25, 64, 67
  Chinese. *See* cilantro
  container depth and, 62
  curled-leaf, 56
  described, **17**, 29, 56, 69
  flat-leaf, 34, **37**, 46, 49–56
  to grow, 53
  Japanese. *See* mitsuba
Parsnip, 83
Path
  herbs for, 51
  permanent, 6
  size of, 6
  weedproof, 6, *80–81*, 83
  width of, 59
Patio, garden on, *10*, 36, 57–61
Pavers, for patio garden, 59
Peach tree, 48
Peanuts, *10*
Peas
  A-frame for, 23
  container depth and, 62
  edible-pod. *See* peas, snow
  flowers, 27
  garden, Little Marvel, 14, **17**
  snap
    'Sugar Ann', **71**
    'Sugar Bon', 14, **17**, 25, 29
    'Sugar Daddy', **19**
  snow (edible-pod pod), 42, 44
    'Oregon Sugar Pod II', 42, **45**
  southern, 44
  supporting, 88–89, *88*
Peat moss, 82, 86
Peat pots, 85

Pepper
to grow, 11
container depth and, 62
hot
'Caliente Hybrid', 14, **17**, 50
'Cayenne', 51, **54**
'Early Jalapeño', 46, **50**
'Jalapeño', **49**
'Riot', **71**
'Serrano', **50**
'Tam Jalapeño', **50**
Thai Chili (Lo Chaio), 42, **45**
semihot
'Anaheim', **50**
'NuMex Joe Parker', *50*
poblano/ancho, 46, **49**
sweet
'Bell Boy Hybrid', 25, 57, 64
described, **29**, 39, 61, 69
'Big Bertha Hybrid', **74**
'Corno di Toro', 31, **32**, 34, 37
'Jingle Bells', 67, **69**
'North Star Hybrid', 14, **17**, 30
'Peperoncino' ('Peperoncini'), **41**
'Purple Beauty', **79**
'Sweet Pickle', 67, **69**
Peppergrass, 54
see also cress, garden
Perennials, to plant, 52–53
Persimmon tree, 48
Pesticides, cautions against, 27
Pe tsai. *See* cabbage, Chinese
Pinks, cottage, 62, **65**, 67
Pipian. *See* squash, 'Zahra Hybrid'
Plan, garden
Asian, *42*
balcony, *67*
basic, 9, *14*, *18*
container, *64*, *67*
to customize, 13
deck, *64*
Plan, garden (*continued*)
"gee whiz," *72*
Italian cooking, *34*, *39*
Mexican, *46*
ornamental/edible, *67*
patio, *57*
salad, *25*, *31*
Planting
intensive, *4–5*, *80–81*
perennials, 52–53
procedures for, 83–86
staggering, 13, 15
Plant lists, 10–11
Asian garden, 44–45
basic garden, 15, 17–19
children's gardens, 24
container gardens, 64–71
"gee whiz" gardens, 75–79
herb gardens, 54–56
Italian gardens, 36–41
Mexican gardens, 48
Native American gardens, 24
patio gardens, 59–61
Plants
to buy, 13
list of. *See* plant lists
Plastic, for mulching, 86, *86*
Plunging, 52–53, 56
Popcorn
'Calico Miniature', 72, 75, **76**
'Mini-Blue', **78**
Portulaca oleracea. See purslane
Potato, 83
new
'All Blue', **33**, 72, 77
'Russian Banana', 11, 14, **17**, *20*, 24, 31, 32
'Yellow Finn', **33**
planting in mulch, 22
'Red Pontiac', 46, **49**
'Russet Burbank', 18, **19**
'Yukon Gold', 72, **77**
Pots
hanging, 64
terra cotta, plunging, 52–53, 56
Potting mix, 62
Pumpkin
'Dill's Atlantic Giant' ('Atlantic Giant PVP'), **74**
flowers, 27
'Jack Be Little', 72, 75, **77**
Japanese, 42, **45**
'Lumina', **79**
signed, 23
'Spirit Hybrid', **17**
'Triple Treat', **24**
Purslane (*verdolaga*) (*Portulaca oleracea*), 46, 47, **49**
to cook, 48

Q, R
Quelite, 46, 47, **49**
Radicchio, 'Medusa Hybrid', 25, **29**, 34, 37
Radish, 62, 83, 85
'Cherry Belle', 14, *20*, 46, 57, 64, 67
described, **17**, 24, 49, 61, 69
'French Breakfast', 25, 34, 67
described, **29**, 37, 69
Long Black Spanish, 72, 73, 74, **77**
'Mino Early Daikon', 42, **45**
Misato Rose Flesh, 72, **78**
'Purple Plum', 25, **29**
'Sakurajima Mammoth', **74**
Rapini. *See* broccoli raab
Rhubarb, *10*, 13, 57, **61**, 83
Rocket. *See* arugula
Rooftop, weight limitations of, 62
Roquette. *See* arugula
Rose, 27
Rosemary, 34, **38**, 51, 56
'Arp', 56
Rucola. *See* arugula

S
Sage, 27
'Aurea', 56
flowers, 59
garden, 51, **56**, 57, 61, 67, 69
'Holt's Mammoth', 56
pineapple, 57, **61**
'Tricolor', 56
Salad, garden for, 9, 25–33
Salad burnet, 51, **56**
Saladini. *See* mesclun
Salsify, 83
Savory
perennial (winter), 56
summer, 51, **56**
Sawdust, 82
Scallion, 25, 34, 46, 57, 62, 64, 67
described, **29**, 37, 49, 61, 69
Seedbed, to prepare, 85
Seeds, *13*
to buy, 13
by mail-order, 13, 90
experiments with, for children, 23–24
to harvest, 53
to plant, 53, 83, *84*, 85, *85*
Shade, vegetables for, 83
Shungiku. *See* chrysanthemum, garland
Site, for garden, 36, 82
Sling, for melons, 89
Snap pea. *See* peas, snap
Snow pea. *See* peas, snow (edible-pod)
Sod cutter, 82, *82*
Soil
amending, 82–83
clay, 83
pH value of, 83
preparation, 7, 82–83
see also potting mix
Sorrel, French, 31, **32**, 41
Spinach, 83
'Melody Hybrid', 14, **17**
'Nordic Hybrid', 25, **29**
see also Malabar spinach
Sprinklers, 87

Squash
acorn, support for, 89
blossoms, 27
container depth and, 62
growing in a bottle, 23
'Hopi Vatnga', **24**
spaghetti, 75
'Tivoli Bush Hybrid', 72, **78**
summer
'Ghada', 40
'Kuta Hybrid', **40**
'Yellow Bush Scallop', 14, **17**
'Zahra Hybrid' (pipian), 40, **50**
see also zucchini
supporting, 89
winter
'Blue Squash', **74**
'Buttercup', *14*, **18**
'Kabocha', 45
'Kikuza', **45**
'Waltham Butternut', *18*, **19**
see also pumpkin, Japanese
Straw, 82
Strawberry, 57, **61**, 62
Alpine, 64
'Alexandria', 69
'Ruegen Improved', 67, **69**
Sulfur, soil, 83
Sunflower, 59
as bean poles, 22
'Giant', 72, **74**, 78
'Mammoth' ('Mammoth Russian'), *20*, *24*, 57, **61**, 74
Sweet cicely, 51, **56**
Sweet pepper. *See* pepper
Sweet woodruff, 51, **56**
Swiss chard, 62, 83
'Argentata', **41**, 70
'Fordhook Giant', 14, **17**, 46, 48
'Ruby' ('Rhubarb'), **41**

T

Tagetes tenuifolia. See marigold, Signet
Tah tsai (tatsoi), 42, 44, **45**
Tarragon, French, 51, **56**
Tatsoi. *See* tah tsai
Tepee, bean, 23
Thyme, 27, **33**
creeping (*Thymus praecox arcticus*), 51, **56**
English, 31, 34, 38, 46, 49, 56
French, 56
garden (*Thymus vulgaris*), 51, **56**, 67, **69**
to grow, 53
lime, 56
variegated, 56
woolly (*Thymus pseudolanuginosus*), 51, **56**
Thymus
praecox
'Albus', 56
arcticus. *See* thyme, creeping
pseudolanuginosus. See thyme, woolly
vulgaris. See thyme, garden
Tiller, rotary, 82
Tomatillo (husk tomato), *10*, 46, 48, **49**
'Indian Strain', 49
'Purple de Milpa', 49
Tomato, 19, 73
in containers, 62, 63
dwarf indeterminate, 57, 59, 63
to grow, 62, 88
husk. *See* tomatillo
large, 74–75
'Better Boy' VFN Hybrid, 14, **18**, 34, 38, 64
'Better Bush Improved', 57, **61**, 67, 69
'Burpee's Supersteak' VFN Hybrid, **33**
'Delicious', **74**, 75
'Early Girl' VFF Hybrid, **33**
Tomato, large (*continued*)
'Giant Belgium', 72, 75, **78**
'Husky Gold', 57, **61**, 69
'Husky Pink VF Hybrid', **61**, 64, 67
'Park's Whopper Improved' VFFNT Hybrid, 31, **32**
'Pineapple', 72, 75, **78**
Paste
San Marzano, 34, **38**, **50**
'La Padino', 38
'Viva Italia' VFFNA Hybrid, *18*, **19**, 31, 33, 40, 46, 49
"Patio" type, 64
small
'Floragold Basket', 64, 67, **69**
'Golden Pear', *10*
'Gold Nugget', **61**
'Micro-Tom', **71**
'Principe Borghese', 34, **38**
'Red Robin', **71**
'Small Fry VFN Hybrid', **71**
'Sun Gold Hybrid', **30**
'Supersweet 100', **29**
'Sweet 100', 14, **18**, 25, 29
'Toy Boy VF Hybrid', 64, **69**
'Whippersnapper', 57, **61**
'Yellow Canary Hybrid', **71**
'Yellow Pear', 25, **29**
to support, 88, 89, *89*
'Eva Purple Ball', 72, **78**
'Evergreen', 72, 73, **78**
'White Beauty', **79**
yellow husk. *See* ground cherry
'Yellow Stuffer', 72, 73, 74, **78**
Transplants, 83, 85
Trellis
in container, *63*
vegetables on, 44, 89
Turnip, 83
white Japanese, 42, **45**
'Hakurei', 45
'Tokyo Cross', 45
'Tokyo Market', 45

V

Vegetables
big, 74
colors in, 72–73
heirloom, 24, 36, 40, 41, 48, 61, 70, 71, 77
shade-tolerant, 83
unusual, 10, *10*, 72–79
Verdolaga. See purslane
Vines, to support, 88–89
Violet, 27

W–Z

Watercress, 54
Watering, 86–87
automatic, 6–7
Watermelon, *12*
'Carolina Cross', **74**
'Garden Baby Hybrid', 14, **18**
'Sugar Baby', 46, **49**
support for, 89
Weather, and planting, 11
Weed prevention, 23
Zucchini
'Burpee Hybrid', 57, **61**
'Cocozelle', **40**
'Fiorentino', 40
'Green Magic', 14, **17**
'Green Magic II', **71**
'Greyzini' ('Grey Zucchini'), 46, **49**
'Roly Poly', 38
'Ronde de Nice' ('Round French'), 34, **38**, 46, 49, 57, 61
'Tromboncino' ('Zucchetta Rampicante'), 34, 36, **38**